W0050108

SPRINGER TRACTS
IN MODERN PHYSICS

Ergebnisse
der exakten Natur-
wissenschaften

Volume **72**

Editor: G. Höhler

Associate Editor: E. A. Niekisch

Editorial Board: S. Flügge J. Hamilton F. Hund
H. Lehmann G. Leibfried W. Paul

Springer-Verlag Berlin Heidelberg GmbH 1974

Manuscripts for publication should be addressed to:

G. HÖHLER, Institut für Theoretische Kernphysik der Universität, 75 Karlsruhe 1, Postfach 6380

Proofs and all correspondence concerning papers in the process of publication should be addressed to:

E. A. NIEKISCH, Institut für Grenzflächenforschung und Vakuumphysik der Kernforschungsanlage Jülich, 517 Jülich, Postfach 365

ISBN 978-3-662-15550-9   ISBN 978-3-540-38332-1 (eBook)
DOI 10.1007/978-3-540-38332-1

Theory of Van der Waals Attraction

DIETER LANGBEIN

Contents

1. Introduction

1.1. Fields of Application

The theory of van der Waals attraction has a broad variety of applications. There are, first of all, the van der Waals gases, to which the name was first applied. The well-known van der Waals equation of state

$$(p + a/v^2)(v - b) = RT \tag{1.1}$$

tells us that the individual gas molecules do not have access to the total volume v, since the remaining molecules already occupy volume b. The correction term, a/v^2 at the pressure p, implies that the kinetic energy with which the molecules hit the volume boundary is lower than their kinetic energy in the interior, owing to the attractive force of the molecules. The attractive correction term in kinetic energy, velocity and pressure is proportional to $1/v^2$, i.e. proportional to $1/r^6$ if r is the mean separation of the gas molecules.

The force responsible for the attraction between any two molecules is at the same time the origin of condensation and crystallization. There is a large class of van der Waals crystals. The simplest of these are the inert gas crystals, which are composed of spherical atoms. Solid helium normally shows a hexagonal close-packed structure. In addition, a body-centered cubic phase and a face-centered cubic phase are known to exist at high pressures. Solid neon exhibits a face-centered cubic structure, but there is also a hexagonal close-packed phase. All other inert gases crystallize in the face-centered cubic structure.

Both the hexagonal close-packed and the face-centered cubic structure can be built up by stacking two-dimensional hexagonal layers of atoms on top of one another. The difference in the binding energy of the two structures arises from the difference in the van der Waals attraction between next-nearest layers. This attraction would favor the hexagonal close-packed structure if pair interactions are considered only. An energy gain of the face-centered cubic structure may be attributed to the fact that the atoms located in neighboring layers are arranged on straight lines, whereas the hexagonal close-packed lattice has only angled arrays of atoms located in neighboring layers. The multiplet contributions caused by straight atomic arrays are generally larger than those caused by angled arrays.

Attempts have also been made to calculate the binding energy of molecular crystals. Two different lines of approach have been used. The first is to integrate the energy between any two atoms. The second replaces the molecules by geometrically simple bodies such as spheres, cylinders or films and integrates the energy of attraction between them. Both approaches are at best able to suggest the correct order of magnitude for the binding energy. They do not permit detailed investigations into the difference in binding energy of different lattice structures. It is the angular variation of the pair energy terms and the multiplet contributions which critically affect the most favorable structure. Explicit calculations on aromatic hydrocarbons and crystalline polymers have been reported. In both cases H–H interactions have been found to be of prime importance.

Van der Waals attraction further provides the basis of most structural and energetic effects in colloid chemistry and biology. Here again it proves very useful to consider the attraction between geometrically simple structures such as spheres, cylinders or films. Spherical shapes are an obvious model for the attraction of colloidal particles and biological cell-cell attraction. The electric double layer formed by the charging of colloidal particles changes the dielectric properties of the surrounding solvent. The resulting change in van der Waals attraction may be of much longer range and may affect the equilibrium position more strongly than the initiating electrostatic charge. This applies in particular to polar solvents like water, whose dielectric constant in strong electric fields drops from eighty times to six times the vacuum dielectric constant.

Striking examples are the formation of liquid crystalline phases in solutions of water and ampholytic substances and the anomalous swelling of clay soils on addition of water. Many important biological questions, including viral self-assembly, the energetics of muscle and the arrangement and mutual recognition of nucleic and proteic acids, require intensive investigations into the van der Waals attraction between cylinders. Knowledge of the free energies of multilayers is necessary for an understanding of a variety of chemical and biological structures ranging from laminar liquid crystals and myelin nerve sheaths to cell membrane adhesion.

Another very important field of van der Waals attraction is surface physics and chemistry. The physisorption of atoms, molecules, or small particles on solid surfaces, their migration due to variations of the surface properties, and the change in attraction between different particles caused by surface and bulk modes are fundamental problems underlying the vast field of catalysis. In particular, collective plasma modes in conductors give rise to favorable steric configurations by increasing the attraction between adsorbed particles, thus lowering the activation energy of numerous reactions.

Let us finally refer to the large field of adhesion and its numerous implications on washing, dyeing, pouring and sliding, wetting, lubrication, powder compaction and pulverizing, and cold welding. It is a basic requirement of detergents that their dielectric properties diminish van der Waals attraction between clothing and dirt. Similarly, the reason why pentane wets water while octane does not is due to small changes in van der Waals attraction. Several precise measurements on the attraction between cylinders or films at separations ranging from 2–20,000 nm have been reported. It is generally found that the adhesion caused by adsorbed layers predominates over that caused by the bulk material at separations smaller than the thickness of the layer. At separations above 100 nm, retardation effects have to be taken into account. Accurate measurements at separations below 2 nm and direct adhesion measurements are generally hampered by the roughness of available surfaces.

1.2. Intermolecular Potentials

The term of van der Waals attraction generally implies that we are dealing with particles whose separation is large enough to exclude overlap of electronic orbitals. Separations involving overlap are tackled only after the theory for large separations is sufficiently well founded. This must be taken into consideration when applying our findings to van der Waals crystals. Their stability depends on the equilibrium between electronic van der Waals attraction and nucleonic Coulomb repulsion. The breakdown of the electronic screening involves a sub-

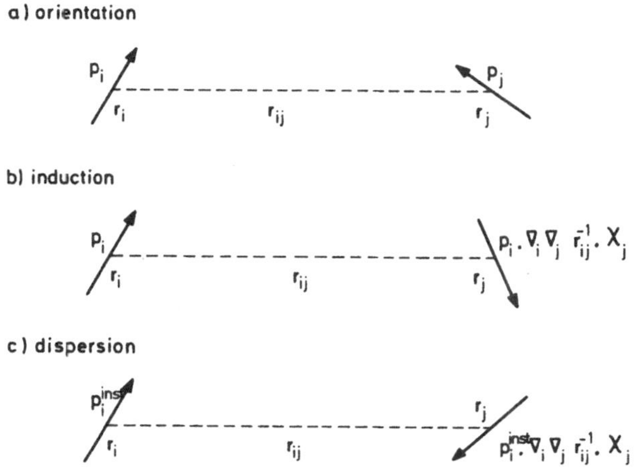

a) orientation

b) induction

c) dispersion

Fig. 1. Dipole interactions

stantial distortion of electron orbitals and considerable electronic overlap. We may expect reliable results on the free energies of van der Waals crystals only if we replace all overlap and exchange interactions by a pseudo-potential for the single electrons. This is a common assumption in solid state theory.

In the range of large separations we distinguish three types of van der Waals attraction, depending on whether the interacting molecules exhibit permanent dipoles or not. Let us consider two molecules i and j with permanent dipoles p_i, p_j at positions r_i and r_j, as shown in Fig. 1a. The mutual interaction potential, which tends to align both molecules, is given by

$$V_{or}(r_{ij}) = p_i \cdot \boldsymbol{V}_i \boldsymbol{V}_j |r_i - r_j|^{-1} \cdot p_j . \qquad (1.2)$$

The orientation giving the lowest energy is the parallel alignment of the two molecules along the joining vector $r_{ij} = r_i - r_j$. This orientation is adopted at very low temperatures, when the molecules are nearly at rest. Hence,

$$\varDelta E_{or} = - p_i p_j [\partial^2 r^{-1}/\partial r^2]_{r_{ij}} = - 2 p_i p_j / r_{ij}^3 \qquad (1.3)$$

for $kT \ll p_i p_j / r_{ij}^3$. At the opposite extreme, when the mean thermal energy kT exceeds the alignment energy $p_i p_j / r_{ij}^3$, the two molecules may rotate and oscillate with high intensity. Orientations with low energy are favored relative to oscillations with high energy according to Boltzmann statistics. Averaging $V_{or} \exp(-V_{or}/kT)$ over all orientations of p_i and p_j yields

$$\Delta E_{or} = -(p_i^2 p_j^2/kT) \left[\mathrm{tr}(\mathbf{\nabla \nabla} 1/r)^2 \right]_{r_{ij}} = -\tfrac{2}{3}(p_i^2 p_j^2/r_{ij}^6)(1/kT) \qquad (1.4)$$

for $kT \gg p_i p_j/r_{ij}^3$. At intermediate temperatures we find that ΔE_{or} depends on the moment of inertia of the molecules under consideration. This orientation effect was first pointed out by Keesom [19, 20]. It provides a $1/r_{ij}^6$ dependence of the free energy ΔE_{or} of orientation on the separation r_{ij}. Nevertheless, the orientation effect fails to explain the attractive term a/v^2 in the van der Waals equation of state (1.1) because it decreases with increasing temperature. Even worse, it disappears between all molecules that do not have permanent dipole moments. Attempts to remove the latter difficulty by considering permanent quadrupoles or octupoles fail likewise to explain van der Waals attraction; any alignment present at low temperatures vanishes at high temperatures. Moreover, the rare gas atoms show van der Waals attraction in spite of their perfectly spherical shape.

While attempting to deduce an interaction which does not vanish with increasing temperature, Debye pointed out that molecules having permanent dipole moments not only align but also polarize each other [21]. An induced dipole rotates simultaneously with the inducing dipole, so that a temperature independent energy gain results. If molecule i exhibits the permanent dipole p_i, the dipole induced at molecule j is given by $p_i \cdot \mathbf{\nabla}_i \mathbf{\nabla}_j r_{ij}^{-1} \cdot \mathbf{X}_j$, and the interaction potential of the two dipoles equals (Fig. 1b)

$$V_{ind}(r_{ij}) = -\tfrac{1}{2} p_i \cdot \mathbf{\nabla}_i \mathbf{\nabla}_j r_{ij}^{-1} \cdot \mathbf{X}_j \cdot \mathbf{\nabla}_j \mathbf{\nabla}_i r_{ij}^{-1} \cdot p_i . \qquad (1.5)$$

The factor $\tfrac{1}{2}$ in (1.5) arises because dipole j is induced by dipole i. $\mathbf{X}_j$ is the polarizability of molecule j. If molecule j also exhibits a permanent dipole p_j, the potential given by Eq. (1.5), and its analogue resulting from interchanging molecules i and j, have to be added to the orientation potential (1.2).

It is again appropriate to distinguish between interactions at low temperatures, where the permanent and the induced dipoles align along the joining vector r_{ij}, to give

$$\Delta E_{ind} = -2(X_j p_i^2 + X_i p_j^2)/r_{ij}^6 \qquad (1.6)$$

and those at high temperatures, where all orientations contribute according to Boltzmann statistics. Hence,

$$\Delta E_{ind} = -(X_j p_i^2 + X_i p_j^2)/r_{ij}^6 . \qquad (1.7)$$

This induction effect yields an attraction proportional to $1/r_{ij}^6$ which decreases, but does not vanish, with increasing temperature. On the other hand, it still requires a permanent dipole moment of at least one molecule, and thus cannot explain the general attraction between arbitrary atoms or molecules.

The true origin of van der Waals attraction is based on quantum mechanics. Quantum theory in its simplest form tells us that everywhere in space there is Planck's quantized radiation field. Everywhere in space photons are moving randomly. These photons are constantly scattered by any particles which are present, so that instantaneous induced dipoles are formed. Each instantaneous dipole p_i^{inst} of molecule i induces a dipole p_j^{ind} of molecule j, which in turn lowers the energy of the instantaneous dipole i, see Fig. 1c. The interaction potential of both molecules is obtained by substituting the instantaneous polarization p_i^{inst} for p_i into Eq. (1.5) and averaging over the time,

$$V_{dis}(r_{ij}) = -\tfrac{1}{2} \langle p_i^{inst} \cdot \boldsymbol{V}_i \boldsymbol{V}_j r_{ij}^{-1} \cdot \mathbf{X}_j \cdot \boldsymbol{V}_j \boldsymbol{V}_i r_{ij}^{-1} \cdot p_i^{inst} \rangle_{av} . \qquad (1.8)$$

The problem now is to find the average polarization $\langle p_i^{inst} p_i^{inst} \rangle_{av}$ of molecule i. It is certainly proportional to the number of photons, which is obtained from Planck's distribution. The coupling parameter between photons and molecules is the molecular polarizability $\mathbf{X}_i$. Its real part $\mathbf{X}_i'$ describes the polarization, its imaginary part $\mathbf{X}_i''$ describes the damping, i.e. the energy dissipated from the molecule to the photon field. In thermal equilibrium there is an equivalent flow of energy from the photons to the molecules, which suggests that the mean intensity of polarization of molecules i and j is proportional to the imaginary part of the respective polarizabilities. This is the general statement of the quasi-classical fluctuation-dissipation theorem derived by Callen and Welton [65].

Substituting into Eq. (1.8) and averaging over time, an energy of attraction is obtained which obeys a $1/r_{ij}^6$ relationship at the separation r_{ij}. ΔE_{dis} occurs between any two molecules, it is given in terms of their polarizabilities, and tends to increase rather than decrease with increasing temperature. The relations between the molecular interaction described here and photon scattering and absorption gave rise to the name dispersion effect.

It is this dispersion effect which explains the general additive cohesion between two particles. The orientation effect and the induction effect require an alignment of permanent dipoles along the vector r_{ij} joining the interacting molecules. The orientation effect is not necessarily additive between three molecules and in many cases repulsion of the third molecule rather than attraction is to be expected. Similarly, the induction effect is greatly reduced if many molecules superimpose their polarizing field from different sides. It is for these reasons that the main emphasis has been placed on the dispersion effect since its interpretation by London in 1930 [22–25]. The rapidly increasing information about the frequency dependence of polarizabilities of different molecules enables quantitative investigations into the van der Waals free energy between atoms and molecules and between macroscopic particles to be undertaken.

1.3. Theoretical Approaches

The original explanation of the dispersion effect by London uses slightly more quantum mechanics than that presented in the preceding section. The basis of his theory is the assumption that macroscopically neutral atoms are not neutral microscopically [22–25]. An electron located at a particular atom sees the instantaneous rather than the average position of the electrons at neighboring atoms. It considers the neighboring atoms to be instantaneous dipoles. This yields a correlation of electron orbitals. London solved the many-electron Schrödinger equation by second-order perturbation theory and expressed the energy gain caused by two-electron correlations in terms of one-electron excitations.

This again demonstrates the relationship between van der Waals attraction and photon absorption and emission. The Schrödinger formalism, instead of explicitly considering photons, contains their average electromagnetic field. The correct free energy of the coupled electron-photon system is obtained only if both the electron system and the photon system remain essentially in their ground states. This implies that only direct electron transitions to and from an excited state, but no successive transitions through different excited states, are allowed. Similarly, only zero or one photon of each type is admitted.

Due to the restriction to direct electron transitions to and from an excited state, each transition may be represented by a dipole oscillator. Since only single excitations of these oscillators are admitted, it is permissible to assume them to be harmonic. The interacting atoms are now interacting harmonic dipole oscillators whose characteristic frequencies are equal to the electronic excitation energies. We obtain the classical Drude model of interaction between electromagnetic fields and matter. It is possible to express the van der Waals energy between two atoms or molecules in terms of oscillator strengths. Quantum theory is now taken into account by providing each dipole oscillator with its average quantum energy, i.e. by applying Bose statistics. The total energy of the coupled dipole system drops if the molecules approach each other because each dipole vibrates in correlation with the others.

The resulting energy gain can be represented by an integral over the polarizabilities of the single molecules, or more roughly by their classical oscillator strengths.

The various interpretations of the dispersion effect are aimed at minimizing the quantum theoretical effort and at expressing the resulting energy gain by measurable quantities such as the electric and magnetic susceptibilities of the particles considered. We used the fluctuation-dissipation theorem, the average quantum energy of Bosons, or the Schrödinger formalism and introduced the electric and magnetic susceptibilities macroscopically using Maxwell's equations or microscopically in terms of one-electron excitations.

The exact quantum theoretical treatment of the dispersion effect involves quantizing matter and electromagnetic fields as well. The coupled electron-photon system is to be treated on the basis of quantum electrodynamics. Using the method of second quantization, it is possible to build up the total Hamiltonian from an electron Hamiltonian H_{el}, a photon Hamiltonian H_{ph}, and an electron-photon interaction operator H_{int}. The dispersion energy between two particles now results in fourth order perturbation. Each contribution is due to the interaction of two electrons with two photons.

If the electron system is assumed to be essentially in its ground state and averaged over all instantaneous excitations, we may define electric and magnetic susceptibilities and return to Maxwell's equations. By assuming the photon system to be essentially in its ground state, and averaging over all instantaneous excitations, we may introduce electric and magnetic interaction potentials and return to Schrödinger's equation. In both cases we obtain equations requiring second order perturbation theory.

It is one of the objectives of the quantum electrodynamic procedure to check the errors and limits of these semiclassical approaches. In particular there is the question of statistics. What is the reason for the dominant influence of Bose statistics in the final energy expression? Another question is up to which order of the electron-photon coupling operator do the semiclassical approaches hold? Is one of these approaches more powerful than the other or limited to certain molecules or separations?

The quantum electrodynamic procedure replaces electrons and photons by quasi-particles. The quasi-electrons are renormalized with respect to terms quadratic in the electron-photon coupling operator. They differ from Fermi statistics with respect to fourth order terms. Similarly, the quasi-photons are renormalized with respect to terms quadratic in the electron-photon coupling operator. They differ from Bose statistics with respect to fourth order terms.

An immediate effect of large separations of the interacting particles is retardation. When discussing the different contributions to van der Waals attraction in the preceding section, we introduced electrostatic potentials $V_{or}(r_{ij})$, $V_{ind}(r_{ij})$ and $V_{dis}(r_{ij})$. These potentials do not take into account the propagation time r_{ij}/c which the field of an oscillating or rotating dipole i needs before it reaches and orientates or polarizes molecule j. The reaction of molecule j arrives at molecule i delayed by $2r_{ij}/c$, where c is the velocity of light, giving rise to a weaker correlation and a smaller energy gain than in the case of an immediate reaction. This retardation generally has little influence on the orientation and induction effects, which both require some nucleonic motion. On the other hand, retardation may become very important in the case of the dispersion effect where electron transitions are considered.

The inclusion of retardation is rather obvious in the dipole model of van der Waals attraction. There is no need to use electrostatic potentials, the electrodynamic four potential of an oscillating dipole is readily found by solving Maxwell's equations rather than Laplace's equation.

The situation becomes more difficult if the Schrödinger formalism is adopted. The Schrödinger formalism is non-relativistic, the many-electron Schrödinger equation assumes static electric and magnetic interaction potentials between all electrons. It does not account for the propagation of fields, i.e. for the fact that electrons located at different molecules see their mutual position and momentum rather late. The inclusion of retardation to the Schrödinger formalism is possible by following the concept used in the retarded oscillator model. It is necessary to introduce time-dependent interaction potentials and, consequently, to solve the many-particle Schrödinger equation by time-dependent perturbation theory.

The decrease in electronic correlation energy as a result of retardation depends on the phase shift involved, i.e. the frequency of the oscillator considered multiplied by the retardation $2r_{ij}/c$ of the reaction field. With increasing separation, the correlation of high frequency oscillators vanishes prior to that of low frequency oscillators; finally at large separations the nearly electrostatic interactions are predominant. The findings on the dispersion energy between two particles in the case of retardation generally exhibit an additional factor $1/r_{ij}$ compared with the non-retarded case.

1.4. Schematic Classification

Figure 2 shows a schematic classification and the interrelations of the different approaches.

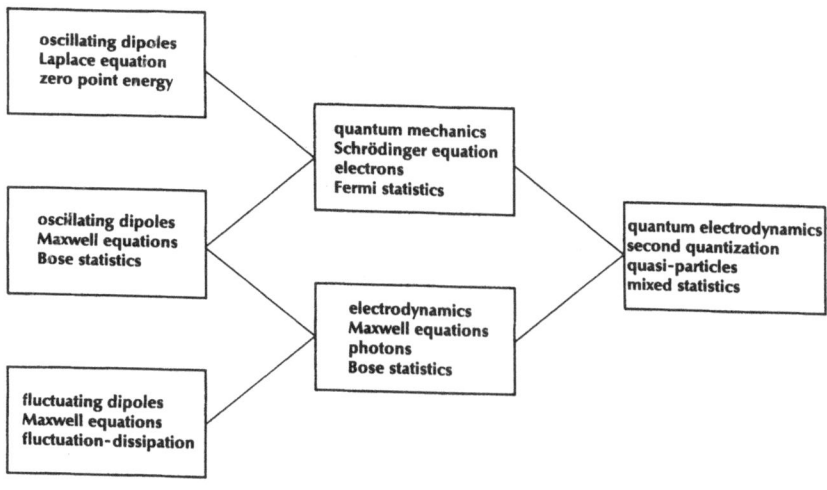

Fig. 2. Schematic classification

In the Drude model of electromagnetic scattering and absorption matter is replaced by harmonic dipole oscillators. Its relevance to the theory of van der Waals attraction was recognized by London who proved on the basis of Schrödinger's equation that the correct interaction energy is obtained if all oscillators have the quantum theoretical zero point energy. When the oscillator model of matter is adopted, there is no reason not to quantize the electromagnetic field. We can understand the zero-point energy as instantaneous electron-photon interactions, i.e. we may consider fluctuating dipoles whose mean intensity is given by the fluctuation-dissipation theorem. It is then possible to account for temperature and retardation as well. The distinction between a fluctuating dipole 1 and an induced dipole 2 is a typical element of perturbation theory. We should rather try to find the exact normal modes of the coupled oscillator system. In this case it is more appropriate to consider the excited oscillator states occupied as required by Bose statistics, rather than to use the fluctuation-dissipation theorem.

These three stages of the oscillator model are shown on the left-hand side of Fig. 2. Their basis is the quantization of matter (electron states) on the one hand, and the quantization of electromagnetic fields (photons) on the other. In the center of Fig. 2 we have placed the Schrödinger equation, from which the dispersion energy is derived using the correlation of electron orbitals and averaging the photon field, and the Maxwell equations, from which the dispersion energy is calculated by providing the classical electromagnetic fields with their average quantum energy.

The exact quantum electrodynamic treatment of the coupled electron-photon system is shown in the right-hand box of Fig. 2.

The approaches discussed so far have been conceived for the treatment of atoms or molecules. These microscopic particles can be replaced with reasonable accuracy by harmonic dipoles. To find the energy between macroscopic particles, all interaction energies between any multiplet of atoms or molecules have to be integrated. The main contribution to the attraction at large separations is due to the additive pair contributions. The triplet and multiplet interactions screen the interaction between inner molecules, whereas they favor that between outer molecules, so that the attractive force is further increased.

For very small separations of the particles or in cases where the binding energy of molecular crystals is considered, a multiplet expansion of the van der Waals energy is no longer possible. In these circumstances we should rather look for the exact normal modes of the coupled oscillator system.

Since Maxwell's equations apply to macroscopic systems, there is no reason for not using the electrodynamic approach discussed previously directly for macroscopic particles. The concept of fluctuations, in fact, was first used by Lifshitz in investigations on the van der Waals attraction between half-spaces [28]. He coupled the electromagnetic fluctuations within the half-spaces to the free electromagnetic modes in the interspace and calculated the change in field energy relative to that in the case of infinite separation. His investigations and the subsequent simplifications reported by van Kampen et al. led to the procedure most widely used at present [35]. Rigorous solutions of Maxwell's equations have been derived for bodies of planar, cylindrical and spherical symmetry [4].

2. Pair Interactions

2.1. Susceptibility

The minimum quantum theoretical effort and the possibility of calculating the van der Waals energy from macroscopic susceptibilities have strongly favored the oscillator model. There are more exact approaches with respect to the theory, which do not, however, aid the interpretation of the experimental data available at present.

Let us consider an ensemble of classical harmonic oscillators $\{i\}$ of vibrating mass m_i, elongation u_i, and frequency ω_i. In order to find their interaction energy we first investigate the reaction of a single oscillator j to an external force F_{ext}. If this force is small enough to ensure linearity, we may write down the Hamiltonian

$$H = \tfrac{1}{2} m_j (\dot{u}_j^2 + \omega_j^2 u_j^2) - F_{ext} u_j. \tag{2.1}$$

Application of the Hamiltonian formalism yields the equation of motion

$$m_j(\ddot{u}_j + \omega_j^2 u_j) = F_{ext} \,. \tag{2.2}$$

If the external force is periodic in time with frequency ω

$$F_{ext} = F(\omega) \exp(-i\omega t) \tag{2.3}$$

it enforces an oscillation of the same frequency ω,

$$u_j = u_j(\omega) \exp(-i\omega t) \,. \tag{2.4}$$

Inserting Eqs. (2.3) and (2.4) into (2.2) we obtain the enforced amplitude

$$u_j(\omega) = F(\omega) \left[m_j(\omega_j^2 - \omega^2) \right]^{-1} \,. \tag{2.5}$$

It is convenient to relate the enforced elongation to the external force by means of a generalized susceptibility

$$u_j = \chi_j(\omega) F_{ext} \tag{2.6}$$

where

$$\chi_j(\omega) = \left[m_j(\omega_j^2 - \omega^2) \right]^{-1} \,. \tag{2.7}$$

If the external force is an arbitrary rather than a periodic function of time, we may represent it by the Fourier integral

$$F_{ext} = \int_{-\infty}^{+\infty} d\omega \, F(\omega) \exp(-i\omega t) \,. \tag{2.8}$$

The enforced elongation then equals

$$u_j = \int_{-\infty}^{+\infty} d\omega \, \chi_j(\omega) \, F(\omega) \exp(-i\omega t) \,. \tag{2.9}$$

The above relations are independent of the type of oscillators considered. They apply to molecular dipole or multipole oscillators and to electromagnetic waves as well.

2.2. Interaction Energy

Let us now turn to two coupled oscillators i and j. An elongation of oscillator i gives rise to an external force F_{ij} on oscillator j. Assuming linear response, we put

$$F_{ij} = u_i T_{ij}(\omega) \tag{2.10}$$

i.e. the equations of motion of the coupled oscillators i and j are

$$\left.\begin{aligned} m_i(\ddot{u}_i + \omega_i^2 u_i) &= u_j\, T_{ji}(\omega) \\ m_j(\ddot{u}_j + \omega_j^2 u_j) &= u_i\, T_{ij}(\omega) \end{aligned}\right\} \tag{2.11}$$

where

$$T_{ij}(\omega) = T_{ji}(\omega) \tag{2.12}$$

according to the action = reaction principle.

By considering only two coupled oscillators i and j we are able to calculate the normal modes explicitly. Introducing

$$u_i = u_i(\omega)\exp(-i\omega t); \qquad u_j = u_j(\omega)\exp(-i\omega t) \tag{2.13}$$

we find from (2.11)

$$\begin{vmatrix} m_i(\omega^2 - \omega_i^2) & T_{ji}(\omega) \\ T_{ij}(\omega) & m_j(\omega^2 - \omega_j^2) \end{vmatrix} = 0 \tag{2.14}$$

and

$$(\omega^2 - \omega_i^2)(\omega^2 - \omega_j^2) = m_i^{-1} T_{ij} m_j^{-1} T_{ji}. \tag{2.15}$$

If the force coefficients $T_{ij}(\omega)$, $T_{ji}(\omega)$ do not depend on frequency, we obtain two normal modes with eigenfrequencies

$$\Omega_{1,2} = \left[\tfrac{1}{2}(\omega_i^2 + \omega_j^2) \pm \left[\tfrac{1}{4}(\omega_i^2 - \omega_j^2)^2 + m_i^{-1} T_{ij} m_j^{-1} T_{ji}\right]^{\frac{1}{2}}\right]^{\frac{1}{2}}. \tag{2.16}$$

In order to obtain the energy of interaction between the two oscillators i and j, we make use of the fact that the quantum energy of a harmonic oscillator with frequency ω and occupation number n_ω equals $\hbar\omega(n_\omega + \tfrac{1}{2})$. The ground state energy of the coupled oscillator system differs from that of the uncoupled oscillator system by

$$\Delta E = \tfrac{1}{2}\hbar(\Omega_1 + \Omega_2) - \tfrac{1}{2}\hbar(\omega_i + \omega_j). \tag{2.17}$$

Inserting (2.16) and expanding the resulting expression in terms of $T_{ij} T_{ji}/m_i m_j$, we find

$$\Delta E = -\tfrac{1}{4}\hbar m_i^{-1} T_{ij} m_j^{-1} T_{ji}\left[\omega_i \omega_j(\omega_i + \omega_j)\right]^{-1}. \tag{2.18}$$

To simplify Eq. (2.18), we use the definite integral

$$\left[\omega_i \omega_j(\omega_i + \omega_j)\right]^{-1} = \pi^{-1}\int\limits_{-\infty}^{+\infty} d\omega \left[(\omega^2 + \omega_i^2)(\omega^2 + \omega_j^2)\right]^{-1} \tag{2.19}$$

yielding

$$\Delta E = -(\hbar/4\pi)\int\limits_{-\infty}^{+\infty} d\omega\, \chi_i(i\omega)\, T_{ij}\chi_j(i\omega)\, T_{ji}. \tag{2.20}$$

This is the simplest form of the susceptibility formula which was derived by McLachlan in 1962 [61, 62].

It is possible to represent the energy of interaction between oscillators of any kind by an integral over their susceptibilities multiplied by their interaction fields along the imaginary frequency axis.

2.3. Electric Dipoles

The susceptibility formula (2.20) can readily be used for calculating the attraction between atoms or molecules, if these particles are replaced by electric dipole oscillators. Let us consider two electric dipole oscillators i and j with elongations u_i and u_j at positions r_i and r_j. The electrostatic force exerted on dipole j by a unit moment of dipole i is given by the respective component of the dipole interaction tensor

$$\mathbf{T}_{ij} = - V_i V_j |r_i - r_j|^{-1} . \tag{2.21}$$

Inserting Eq. (2.21) into (2.20) yields the energy of interaction between dipoles i and j.

In order to describe the behavior of a molecule i we need three-dimensional dipole oscillators, each being equivalent to three independent oscillators vibrating in orthogonal directions. The total energy of interaction between the orthogonal oscillators representing molecule i and the orthogonal oscillators representing molecule j is obtained by summing Eq. (2.20) over all space directions. Allowing for different susceptibilities of molecules i and j in orthogonal space directions, i.e. introducing susceptibility tensors rather than scalars gives

$$\Delta E = - (\hbar/4\pi) \int_{-\infty}^{+\infty} d\omega \, \text{tr} \, \{ \boldsymbol{\chi}_i(i\omega) \cdot \mathbf{T}_{ij} \cdot \boldsymbol{\chi}_j(i\omega) \cdot \mathbf{T}_{ji} \} . \tag{2.22}$$

The semiclassical substitution of electric dipole oscillators for molecules requires that one three-dimensional dipole is attached to each allowed electron transition. Each molecule has to be replaced by an ensemble $\{k\}$ of independent three-dimensional dipole oscillators whose eigen-frequencies correspond to the allowed one-electron excitation energies. The energy of interaction of molecules i and j is derived by summing Eq. (2.22) over all dipoles $\{k\}$ representing molecule i and over all dipoles $\{l\}$ representing molecule j. These summations do not affect the dipole interaction tensors $\mathbf{T}_{ij} \mathbf{T}_{ji}$, which depend exclusively on the separation and mutual orientation of molecules i and j. We are left with the summation over the susceptibilities $\chi_k(\omega)$ of dipoles k at molecule i

and with that over the susceptibilities $\chi_l(\omega)$ of dipoles l at molecule j. If we now introduce molecular polarizabilities,

$$
\mathbf{X}_i(\omega) = \sum_{k \in i} \boldsymbol{\chi}_k(\omega) = \sum_{k \in i} \mathbf{m}_k^{-1}(\omega_k^2 - \omega^2)^{-1} \, ;
$$

$$
\mathbf{X}_j(\omega) = \sum_{l \in j} \boldsymbol{\chi}_l(\omega) = \sum_{l \in j} \mathbf{m}_l^{-1}(\omega_l^2 - \omega^2)^{-1}
$$

$$(2.23)$$

we obtain

$$
\Delta E = -(\hbar/4\pi) \int_{-\infty}^{+\infty} d\omega \, \mathrm{tr} \, \{ \mathbf{X}_i(i\omega) \cdot \boldsymbol{V}_i \boldsymbol{V}_j r_{ij}^{-1} \cdot \mathbf{X}_j(i\omega) \cdot \boldsymbol{V}_j \boldsymbol{V}_i r_{ij}^{-1} \} \, .
$$

$$(2.24)$$

It is possible to find the interaction energy between two molecules by integrating their polarizabilities multiplied by the dipole interaction tensor along the imaginary frequency axis. Since in Eq. (2.24) we use the electrostatic interaction tensor $\mathbf{T}_{ij}$, the frequency integration actually affects only the polarizabilities. These quantities are strictly real on the imaginary frequency axis.

The dipole interaction tensor (2.21) is obviously proportional to the inverse cube of the separation $r_{ij} = |\mathbf{r}_i - \mathbf{r}_j|$ of molecules i and j. The interaction energy ΔE is found to be proportional to the inverse sixth power of r_{ij}. In the simplest case of spherical atoms, we obtain

$$
\mathrm{tr} \, \{ \mathbf{T}_{ij} \mathbf{T}_{ji} \} = 6/r_{ij}^6
$$

$$(2.25)$$

and

$$
\Delta E = -(3\hbar/2\pi) \, r_{ij}^{-6} \int_{-\infty}^{+\infty} d\omega \, X_i(i\omega) X_j(i\omega)
$$

$$(2.26)$$

since $X_i(\omega)$ and $X_j(\omega)$ are scalars. This basic $1/r_{ij}^6$ relationship between the dispersion energy ΔE and the separation r_{ij} was first derived by London [24]. London and Eisenschitz considered several dipole oscillators per atom or molecule [22]. The representation of the interaction parameter by a frequency integral over the polarizabilities involved was first used in the investigations on retarded interactions reported by Casimir and Polder [26].

2.4. Spheres

The susceptibility formula Eq. (2.20) can be used to describe the dispersions energy between macroscopic particles 1 and 2 along two alternative routes. We may replace each atom of particles 1 and 2 by a set of dipole oscillators and integrate the dispersion energy between any pair of atoms, or we may treat the particles as a whole as harmonic oscillators.

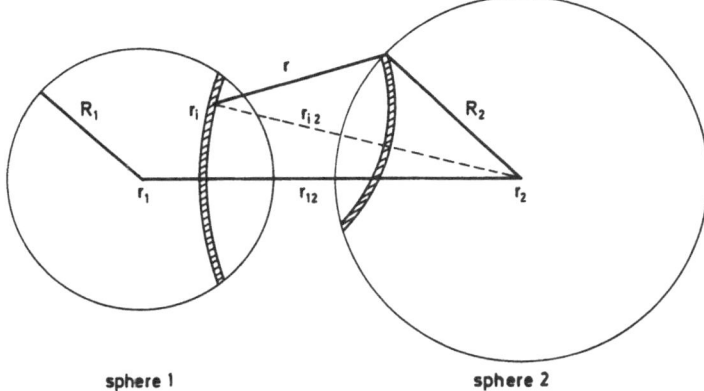

Fig. 3. Attracting spheres

In the latter case we are not allowed to restrict ourselves to dipoles, but have to include 2^n poles of arbitrary order.

In this section we shall consider only the first approach. An exact integration of Eq. (2.26) can be performed if particles 1 and 2 are homogeneous spheres with radii R_1 and R_2. We consider the integral

$$C_m = \int_1 d\mathbf{r}_i \int_2 d\mathbf{r}_j r_{ij}^{-m} \tag{2.27}$$

where m is arbitrary in order to include higher order interactions in the integration. The integral over sphere 2 is conveniently evaluated by introducing spherical coordinates at position $\mathbf{r}_i$ in sphere 1, as shown in Fig. 3.

Hence,

$$C_m = \int_1 d\mathbf{r}_i (\pi/r_{i2}) \int_{r_{i2}-R_2}^{r_{i2}+R_2} dr [R_2^2 - (r_{i2}-r)^2]/r^{m-1} \tag{2.28}$$

and

$$C_m = 2\pi \int_1 d\mathbf{r}_i r_{i2}^{-1} \{R_2(m-2)^{-1}(m-3)^{-1}[(r_{i2}-R_2)^{-(m-3)} + (r_{i2}+R_2)^{-(m-3)}]$$
$$- (m-2)^{-1}(m-3)^{-1}(m-4)^{-1}[(r_{i2}-R_2)^{-(m-4)} - (r_{i2}+R_2)^{-(m-4)}]\}. \tag{2.29}$$

Similarly, for the integration over sphere 1 we use spherical coordinates at the center $\mathbf{r}_2$ of sphere 2 to give

$$\int_1 d\mathbf{r}_i \cdot r_{i2}^{-1} (r_{i2} \pm R_2)^{-(n-1)} = \pi r_{12}^{-1} \int_{r_{12}-R_1}^{r_{12}+R_1} dr \cdot (r \pm R_2)^{-(n-1)} [R_1^2 - (r_{12}-r)^2] \tag{2.30}$$

and

$$\int_1 dr_i \cdot r_{i2}^{-1}(r_{12} \pm R_2)^{-(n-1)}$$

$$= 2\pi r_{12}^{-1} \{ R_1 (n-2)^{-1}(n-3)^{-1} [(r_{12} \pm R_2 - R_1)^{-(n-3)}$$

$$+ (r_{12} \pm R_2 + R_1)^{-(n-3)}] \tag{2.31}$$

$$- (n-2)^{-1}(n-3)^{-1}(n-4)^{-1} [(r_{12} \pm R_2 - R_1)^{-(n-4)}$$

$$- (r_{12} \pm R_2 + R_1)^{-(n-4)}] \}.$$

Inserting Eq. (2.31) into Eq. (2.29) for $n = m - 2$ yields

$$C_m = \frac{(2\pi)^2}{(m-2)! r_{12}} \left[\left[\frac{(m-6)! r_i r_j}{(r_{12} + r_i + r_j)^{m-5}} + \frac{(m-7)! (r_i + r_j)}{(r_{12} + r_i + r_j)^{m-6}} \right. \right.$$

$$\left. \left. + \frac{(m-8)!}{(r_{12} + r_i + r_j)^{m-7}} \right]_{-R_1}^{+R_1} \right]_{-R_2}^{+R_2}. \tag{2.32}$$

The main term in Eq. (2.32) at small separations d is that arising for $r_i = -R_1, r_j = -R_2$

$$C_m \simeq (2\pi)^2 (R_1 R_2/r_{12}) d^{-(m-5)} (m-6)!/(m-2)! \tag{2.33}$$

for $d = r_{12} - R_1 - R_2 \ll R_1, R_2$. Hence,

$$\Delta E_{\text{sph}} = -\tfrac{1}{4}\pi\hbar \cdot R_1 R_2 (R_1 + R_2)^{-1} d^{-1} \int_{-\infty}^{+\infty} d\omega \varrho_1 X_1(i\omega) \varrho_2 X_2(i\omega) \tag{2.34}$$

where ϱ_1, ϱ_2 are the densities of atoms or of dipole oscillators in spheres 1 and 2.

In the case of large separations $r_{12} \gg R_1, R_2$ we conveniently expand Eq. (2.32) with respect to $(r_i + r_j)/r_{12}$. The boundaries $r_i = \pm R_1, r_j = \pm R_2$ entail that only those terms are left which are odd with respect to r_i, r_j, so that

$$C_m \simeq \tfrac{4}{3}\pi R_1^3 \tfrac{4}{3}\pi R_2^3 r_{12}^{-m} \quad \text{for} \quad d \gg R_1, R_2. \tag{2.35}$$

At large separations d we find that the integral over the pair interactions obeys the same power law as the pair interactions and is proportional to the volumes of the particles under consideration. This can be concluded also directly from the definition (2.27). Hence,

$$\Delta E_{\text{sph}} = -\tfrac{8}{3}\pi\hbar (R_1^3 R_2^3/r_{12}^6) \int_{-\infty}^{+\infty} d\omega \varrho_1 X_1(i\omega) \varrho_2 X_2(i\omega). \tag{2.36}$$

The dispersion energy between two spheres 1 and 2 is inversely proportional to their separation for small values of d and to the sixth power of their separation for large d.

When applying Eq. (2.32) to interactions obeying power laws $m \leq 7$, we have to keep in mind that $(n-1)! \int dr \, r^{-n} = \ln r$ rather than $-(n-2)! \, r^{-(n-1)}$ for $n = 1$. However, this necessary change in notation does not affect the conclusions drawn from Eq. (2.32).

2.5. Small Separations

An exact integration of pair interactions is restricted to a few simple structures like spheres or half-spaces. The case of half-spaces was treated by de Boer in 1936 [90], the treatment of spheres was reported one year later by Hamaker [91].

If other structures are considered we have to introduce approximations before carrying out integrations of the type considered in Eq. (2.27). It is often convenient to use different approximations for small and large separations. At small separations d it is appropriate to count the number of possibilities $P(r)$ in which a fixed distance $r = r_{ij}$ can be placed between atoms i of particle 1 and atoms j of particle 2. Between spheres with radii R_1, R_2 we find the weight function

$$P(r) \simeq \tfrac{2}{3}\pi^2 (R_1 R_2 / r_{12}) \, r (r - d)^3 \tag{2.37}$$

which, by insertion into Eq. (2.27), directly yields Eq. (2.33). If ellipsoidal particles with arbitrary curvature are considered, the structure factor in Eq. (2.37) changes correspondingly, whereas the cubic increase of $P(r)$ with $r - d$, and the d^{-1} relationship between dispersion energy and separation are maintained.

The dependence of the weight function $P(r)$ on the distance r changes strongly if parallel cylinders or half-space are considered. In the case of parallel cylinders the integration parallel to the cylinder axis may be carried out prior to that normal to it, so that a two-dimensional weight function $P(r)$ is obtained. In the case of half-spaces, we may first carry out the integrations tangential to their surfaces and then merely need a one-dimensional weight function $P(r)$. Using

$$\int_{-\infty}^{+\infty} dz \cdot (r^2 + z^2)^{-m/2} = \Gamma(\tfrac{1}{2}m - \tfrac{1}{2}) \, \Gamma(\tfrac{1}{2}) / \Gamma(\tfrac{1}{2}m) \, r^{m-1} \tag{2.38}$$

we find for cylinders with length L large compared with their separation d, and their radii R_1, R_2

$$C_m = [\Gamma(\tfrac{1}{2}m - \tfrac{1}{2}) \, \Gamma(\tfrac{1}{2}) / \Gamma(\tfrac{1}{2}m)] \, L \int_1 d\mathbf{r}_i \int_2 d\mathbf{r}_j r_{ij}^{-(m-1)} \tag{2.39}$$

and for half-spaces with surface area L^2

$$C_m = 2\pi(m-2)^{-1} L^2 \int_1 \mathrm{d}r_i \int_2 \mathrm{d}r_j \, r_{ij}^{-(m-2)} . \tag{2.40}$$

The two-dimensional weight function $P(r)$ between cylinders with radii R_1, R_2 is calculated in Section 2.8. We obtain

$$P(r) = 2(R_1 R_2/r_{12})^{\frac{1}{2}} r^{\frac{1}{2}} (r-d)^2 \sum_{n=0}^{\infty} \frac{\Gamma(n+\frac{1}{2})^2}{n!(n+2)!} [(r-d)/2r]^n . \tag{2.41}$$

All powers of $(r-d)/r$ are required by the subsequent integration of Eq. (2.41) from $r=d$ to infinity. The one-dimensional weight function $P(r)$ for half-spaces is readily seen to be

$$P(r) = r - d . \tag{2.42}$$

Substituting (2.41) into Eq. (2.39) and integrating over r gives

$$C_m = 2\pi L(R_2 R_2/r_{12})^{\frac{1}{2}} \frac{\Gamma(m-\frac{9}{2})}{d^{m-9/2}} \frac{\Gamma(\frac{1}{2}m-\frac{1}{2}) \Gamma(\frac{1}{2})}{\Gamma(\frac{1}{2}m) \Gamma(\frac{1}{2}m-\frac{3}{2})} F(\frac{1}{2},\frac{1}{2};m-\frac{3}{2};\frac{1}{2}) . \tag{2.43}$$

The hypergeometric series $F(\frac{1}{2},\frac{1}{2};m-\frac{3}{2};\frac{1}{2})$ can be evaluated rigorously on the basis of Eq. (15.1.26) in Ref. [1]. Hence,

$$C_m = 2\pi(2\pi R_1 R_2/r_{12})^{\frac{1}{2}} [\Gamma(m-\frac{9}{2})/\Gamma(m-1)] L \cdot d^{-(m-9/2)} \tag{2.44}$$

and $(m=6)$

$$\Delta E_{\mathrm{cyl}} = -\tfrac{1}{16}\pi h L d^{-3/2} [2R_1 R_2/(R_1+R_2)]^{\frac{1}{2}} \int_{-\infty}^{+\infty} \mathrm{d}\omega \varrho_1 X_1(i\omega) \varrho_2 X_2(i\omega). \tag{2.45}$$

The dispersion energy between two parallel cylinders of length L obeys a $d^{-3/2}$ relationship for small separations d. By inserting the one-dimensional weight function (2.42) into the interaction integral (2.40) for half-spaces we find straight away

$$C_m = 2\pi [(m-5)!/(m-2)!] L^2 d^{-(m-4)} \tag{2.46}$$

and

$$\Delta E_{\mathrm{hsp}} = -\tfrac{1}{8}\hbar(L/d)^2 \int_{-\infty}^{+\infty} \mathrm{d}\omega \varrho_1 X_1(i\omega) \varrho_2 X_2(i\omega) . \tag{2.47}$$

The dispersion energy between two half-spaces of surface area L^2 is proportional to the inverse square of their separation. The above parallel treatment of interaction integrals between spheres, cylinders and half-spaces shows that the power law to be expected is clearly related to the number of directions in which the particles under consideration exhibit curvatures. If their curvature vanishes in n directions, we may readily carry out the integration over these directions and are then left with the two- or one-dimensional interaction integrals (2.39) and (2.40) and with the two- or one-dimensional weight functions (2.41) and (2.42).

The final Eqs. (2.33), (2.44), and (2.46) can be summarized to yield

$$C_m = 2\pi [2\pi R_1 R_2/(R_1 + R_2)]^{1-n/2}$$
$$\cdot [\Gamma(m + \tfrac{1}{2}n - 5)/\Gamma(m - 1)] L^n d^{-(m + \frac{1}{2}n - 5)} \tag{2.48}$$

n being the number of directions with vanishing curvature,

$$n = \begin{cases} 0 \text{ for ellipsoidal particles} \\ 1 \text{ for cylindrical particles} \\ 2 \text{ for half-spaces} . \end{cases} \tag{2.49}$$

The corresponding Eqs. (2.34), (2.45), and (2.47) for the dispersion energy now read

$$\Delta E = -\tfrac{1}{8}\hbar [2\pi R_1 R_2/(R_1 + R_2)]^{1-n/2}$$
$$\cdot \Gamma(1 + \tfrac{1}{2}n) L^n d^{-(1+\frac{1}{2}n)} \int_{-\infty}^{+\infty} d\omega \varrho_1 X_1 \varrho_2 X_2 . \tag{2.50}$$

Although Eq. (2.50) is obtained by integrating pair interactions only, it yields the correct dependence for the dispersion energy on the separation d, even if higher order interactions are included. The screening caused by the higher order interactions entails that the integral $\int d\omega \varrho_1 X_1 \varrho_2 X_2$ over the polarizabilities of the single atoms has to be replaced by an integral over the susceptibilities of the total particles 1 and 2. However, it does not change the general dependence of the dispersion energy on the separation d.

2.6. Large Separations

Turning to large separations we have to be aware of the fact that our investigations now become more risky than in the case of small separations where the most important contribution of higher order interactions is the screening of pair interactions, which changes the absolute value but not the overall relationship between dispersion energy and separation.

At large separations retardation occurs. Any results derived by integration of pairwise r^{-6} interactions are valid only for separations small enough to exclude large phase shifts of the interacting electromagnetic modes. Since the retardation of all modes on their path from particle 1 to particle 2 and back equals $2r_{12}/c$, we find the phase shift of the high frequency modes to be larger than that of the low frequency modes. The contribution of the high frequency modes to the dispersion

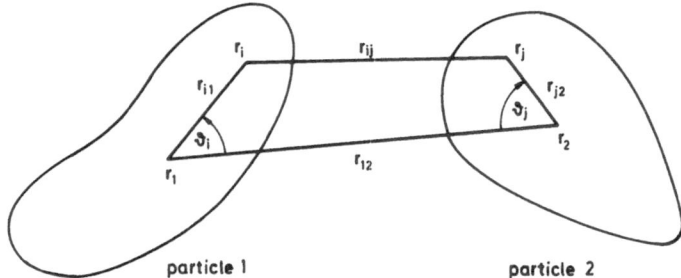

Fig. 4. Large separations

energy vanishes first. We will learn in Chapter 6 that the retarded dispersion energy between small particles follows a d^{-7} relationship rather than a d^{-6} relationship with separation d. Thus, it is again reasonable to integrate pairwise r^{-m} interactions where m is arbitrary.

If the two interacting particles 1 and 2 are finite, their maximum diameters being small compared with their separation, it is permissible to expand the distance r_{ij} between any two atoms of particles 1 and 2 in Taylor series around the centers r_1 and r_2. The centers of particles 1 and 2 are appropriately chosen to be the centers of mass, so that the linear terms in the expansion of r_{ij}^{-m} vanish on integration. Using the notation shown in Fig. 4, we obtain

$$
\begin{aligned}
C_m = r_{12}^{-m} &\int_1 \mathrm{d}r_i \int_2 \mathrm{d}r_j \\
&+ m r_{12}^{-(m+2)} \int_1 \mathrm{d}r_i \int_2 \mathrm{d}r_j \{ r_{i1}^2 (\tfrac{1}{2}(m+2)\cos^2 \vartheta_i - \tfrac{1}{2}) \\
&+ r_{j2}^2 (\tfrac{1}{2}(m+2)\cos^2 \vartheta_j - \tfrac{1}{2}) \} \, .
\end{aligned}
\tag{2.51}
$$

The integral over all pair interactions between the atoms of finite particles is proportional to their volume and obeys essentially the same power law r_{12}^{-m} with the separation r_{12} as do the pair interactions. The higher order terms $r_{12}^{-(m+2)}$ tend to rotate the axes of maximum inertia of particles 1 and 2 to the connecting line r_{12}.

Power laws different from that of the pair interactions arise as soon as we consider particles 1 and 2 to extend infinitely in at least one direction. Let us first consider a small particle 1 and a long thin cylinder 2, see Fig. 5a. Starting with the integration over the cylinder axis analogous to the procedure described in Section 2.5, we find with the help of Eq. (2.38)

$$
C_m = [\Gamma(\tfrac{1}{2}m - \tfrac{1}{2}) \, \Gamma(\tfrac{1}{2}) / \Gamma(\tfrac{1}{2}m)] \int_1 \mathrm{d}r_i \int_{(2)} \mathrm{d}r_j r_{ij}^{-(m-1)}
\tag{2.52}
$$

with the integration over r_j now extending over that cross-section of cylinder 2 which is at a minimum distance from particle 1. The interaction

a) **sphere + cylinder**

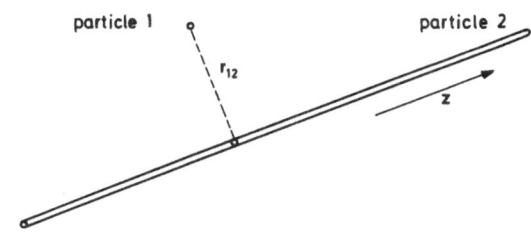

b) **crossed cylinders**

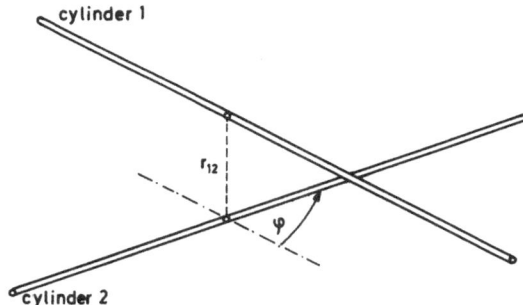

Fig. 5. Extended particles

energy between a small particle and a cylinder varies as $r_{12}^{-(m-1)}$, which is one inverse power less than in the case of two finite particles.

If particle 1 is also a cylinder, the angle included by the cylinder axes being φ, see Fig. 5b, we find by integrating Eq. (2.52) over cylinder 1

$$C_m = [\Gamma(\tfrac{1}{2}m - \tfrac{1}{2})\,\Gamma(\tfrac{1}{2})^2/\sin\varphi\,\Gamma(\tfrac{1}{2}m)]\int_{(1)} d\mathbf{r}_i \int_{(2)} d\mathbf{r}_j r_{ij}^{-(m-2)}. \qquad (2.53)$$

Integration over $\mathbf{r}_i$ extends over the cross-section of minimum distance r_{12} from cylinder 2.

The interaction energy between parallel cylinders is proportional to their length times $r_{12}^{-(m-1)}$, that between crossed cylinders is proportional to $r_{12}^{-(m-2)}$. The extension of particle 1 in an additional direction clearly lowers the dependence of the interaction energy on the separation by another inverse power of r_{12}. This principle holds in a similar manner if particles extended in further directions are considered. If particle 1 is a plane and particle 2 is a parallel thin cylinder, we obtain an interaction energy proportional to the cylinder length times $r_{12}^{-(m-2)}$, that between two parallel planes is proportional to their surface times $r_{12}^{-(m-2)}$. If we consider particles extended in the direction parallel to their separation, we find that the dependence of the interaction energy on the separation is lowered by another inverse power of r_{12}.

Summarizing these findings, we obtain

$$\Delta E \propto r_{12}^{-(m-n)} \tag{2.54}$$

where n is the number of independent directions in which the interacting particles extend. Neglecting retardation, we may put $m = 6$. Anticipating the retarded pair interaction energy discussed in Chapter 6 we obtain $m = 7$.

2.7. Asymptotic Power Laws

Several examples of the asymptotic power laws Eqs. (2.50) and (2.54) at small and large separations are collected in Table 1. Though we integrated pair interactions only in the investigations discussed here, setting up

Table 1. Asymptotic power laws

Separation	Interacting particles	Non-retarded limit	Retarded limit
small separations $d \ll$ charact. diameters	sphere + sphere + cylinder + half-space crossed cylinders	d^{-1}	d^{-2}
	parallel cylinders cylinder + half-space	$d^{-3/2}$	$d^{-5/2}$
	half-space + half-space	d^{-2}	d^{-3}
large separations $d \gg$ charact. diameters	half-space + half-space plate normal to half-space rod normal to half-space crossed plates	d^{-2}	d^{-3}
	film parallel to half-space cylinder parallel to half-space sphere + half-space plate normal to film	d^{-3}	d^{-4}
	parallel films cylinder parallel to film sphere + film crossed cylinders	d^{-4}	d^{-5}
	parallel cylinders sphere + cylinder	d^{-5}	d^{-6}
	sphere + sphere	d^{-6}	d^{-7}

this table is still permissible. The multiplet interactions increase the absolute values, but do not alter the asymptotic power laws of the dispersion energy between interacting particles. At small separations the applicable power law is determined by the number of directions with vanishing curvature, at large separations it depends on the number of directions which show infinite extension of the interacting partners.

2.8. Cylinders, Small Separations

In this section we will calculate the two-dimensional weight function $P(r)$ arising between parallel cylinders. We have to find the number of possibilities for a fixed distance $r = r_{ij}$ to occur between any position r_i within a circle of radius R_1 around r_1 and any position r_j within a circle of radius R_2 around r_2, see Fig. 6a. We obtain the number of possible ways of arranging a given vector r between r_i and r_j by shifting circle 1 by r and calculating the resulting overlap area of both circles, see Fig. 6a.

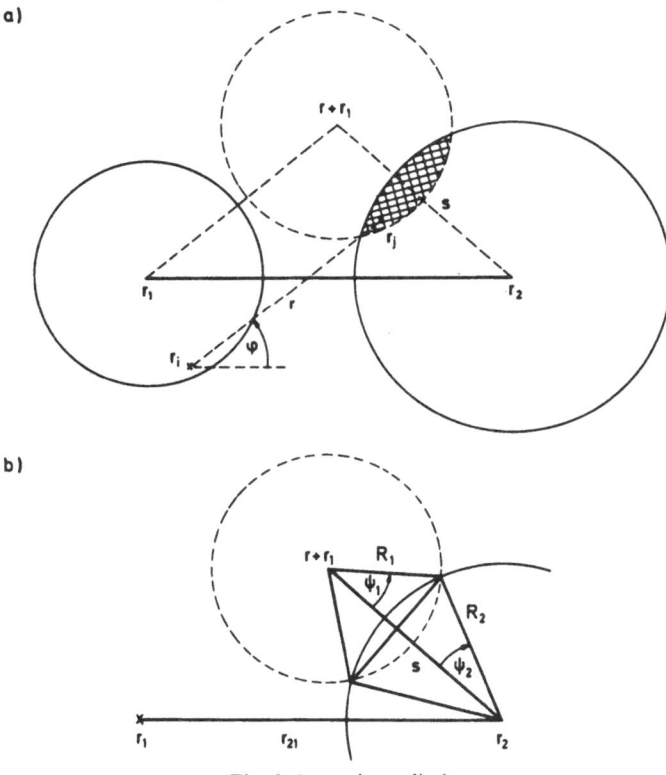

Fig. 6. Attracting cylinders

This overlap area equals

$$\sum_{i=1,2} 2R_i^2 \int_0^{\psi_i} \mathrm{d}\psi \sin^2\psi = \sum_{i=1,2} R_i^2 (\psi_i - \sin\psi_i \cos\psi_i). \tag{2.55}$$

By integrating Eq. (2.55) over all angles φ formed by r and the connecting line $r_{21} = r_2 - r_1$ we obtain

$$P(r) = \sum_{i=1,2} 4r \int_0^{\varphi_0} \mathrm{d}\varphi \, R_i^2 \int_0^{\psi_i} \mathrm{d}\psi \sin^2\psi. \tag{2.56}$$

The maximum angle φ_0 is the angle for which the displaced circle 1 and circle 2 just touch one another. If s denotes the separation of the shifted centers, we learn from Fig. 6b

$$s^2 = r_{12}^2 + r^2 - 2r_{12} r \cos\varphi \tag{2.57}$$

$$2s R_1 \cos\psi_1 = s^2 + R_1^2 - R_2^2; \quad 2s R_2 \cos\psi_2 = s^2 + R_2^2 - R_1^2. \tag{2.58}$$

In Eq. (2.56) we find it convenient to use the left-hand side of Eq. (2.55) rather than the integrated right-hand side. By interchanging the order of integration we obtain

$$P(r) = \sum_{i=1,2} 4r R_i^2 \int_0^{\psi_{i0}} \mathrm{d}\psi \sin^2\psi \, \varphi_i(\psi) \tag{2.59}$$

where $\varphi_1(\psi)$ and $\varphi_2(\psi)$ are the inverse functions of $\psi_1(\varphi)$ and $\psi_2(\varphi)$, respectively. From Eqs. (2.57) and (2.58) we find

$$\varphi_i(\psi) = 2 \arcsin \left(\left[(R_i \cos\psi + \sqrt{R_j^2 - R_i^2 \sin^2\psi})^2 \right. \right.$$
$$\left. \left. - (r_{21} - r)^2 \right] / 4r_{21} r \right)^{\frac{1}{2}} \tag{2.60}$$

where $j = 2$ for $i = 1$, and vice versa. Here and in the following we obtain the equations valid for sphere 2 from those valid for sphere 1 simply by interchanging subscripts 1 and 2.

The upper integration limits ψ_{i0} in Eq. (2.59) are determined by $\varphi_i(\psi_{i0}) = 0$, i.e.

$$\sin^2 \tfrac{1}{2}\psi_{i0} = (r - d) \left[2R_j - (r - d) \right] / 4(r_{21} - r) R_i \equiv x_{i0}^2. \tag{2.61}$$

Substituting generally

$$x = \sin \tfrac{1}{2}\psi \tag{2.62}$$

and expanding at the respective upper limit x_{i0}, we obtain

$$P(r) = 32r \sum_{i=1,2} R_i^2 \int_0^{x_{i0}} \mathrm{d}x \cdot x^2 \sqrt{1 - x^2} \, \varphi_i(x) \tag{2.63}$$

where

$$\varphi_i(x) = 2 \arcsin \left\{ \left(\frac{2 R_i (x_{i0}^2 - x^2)(r_{21} - r)^3}{r_{21} r [(r_{21} - r)^2 + R_j^2 - R_i^2]} \right)^{\frac{1}{2}} \right.$$

$$\left. \cdot \left[1 + \frac{R_i (x_{i0}^2 - x^2)(r_{21} - r)[r_{21} - r)^2 + 3(R_j^2 - R_i^2)]}{[(r_{21} - r)^2 + R_j^2 - R_i^2]^2} + \cdots \right] \right\}. \tag{2.64}$$

In connection with the x integration in Eq. (2.63) we expand also the arc sine in Eq. (2.64). However, we may not cut off this expansion after a finite number of terms in view of the required integration of $P(r)$ over r from d to infinity. Using

$$\int_0^{x_{i0}} dx \cdot x^{2m}(x_{i0}^2 - x^2)^{n-\frac{1}{2}} = x_{i0}^{2m+2n} \frac{\Gamma(m+\frac{1}{2})\,\Gamma(n+\frac{1}{2})}{2\Gamma(m+n+1)} \tag{2.65}$$

we finally obtain

$$P(r) = r \sum_{n=0}^{\infty} \frac{\Gamma(n+\frac{1}{2})^2}{n!(n+2)!} \sum_{i=1,2} \frac{R_i^{\frac{1}{2}}(r-d)^{n+2}[2R_j - (r-d)]^{n+2}(r_{21}-r)^{2n-\frac{1}{2}}}{(2r_{21} r[(r_{21}-r)^2 + R_j^2 - R_i^2])^{n+\frac{1}{2}}}. \tag{2.66}$$

Expanding Eq. (2.66) with respect to $r-d$ and summing over $i = 1, 2$, we are left with the weight function (2.41) used in Section 2.5.

3. Multiplet Interactions

3.1. Fluctuations

The influence of multiplet interactions on the dispersion energy is conveniently demonstrated by applying the perturbation method based on fluctuations. There is a finite probability for each molecule i to fluctuate, i.e. to absorb a photon from Planck's radiation field, thus producing an instantaneous dipole p_i^{inst} at position r_i. This instantaneous dipole gives rise to the field $p_i^{\text{inst}} \cdot \mathbf{T}_{ij}$ at position r_j of molecule j and induces the dipole

$$p_j^{\text{ind}} = p_i^{\text{inst}} \cdot \mathbf{T}_{ij} \cdot \mathbf{X}_j. \tag{3.1}$$

Dipole j, in turn, gives rise to the field $p_j^{\text{ind}} \cdot \mathbf{T}_{ji}$ at position r_i, i.e. the energy of dipole i is changed by an amount

$$\Delta E_i = -\tfrac{1}{2} \langle p_i^{\text{inst}} \cdot \mathbf{T}_{ij} \cdot \mathbf{X}_j \cdot \mathbf{T}_{ji} \cdot p_i^{\text{inst}} \rangle_{\text{av}}. \tag{3.2}$$

The brackets in Eq. (3.2) indicate that we have to take the average over time. Substituting the Fourier representation of p_i^{inst}

$$p_i^{inst}(t) = \int\limits_{-\infty}^{+\infty} d\omega\, p_i(\omega)\exp(-i\omega t) \tag{3.3}$$

we obtain

$$\Delta E_i = -\tfrac{1}{2}\left\langle \int\limits_{-\infty}^{+\infty} d\omega \int\limits_{-\infty}^{+\infty} d\omega'\, p_i(\omega)\cdot \mathbf{T}_{ij}(\omega)\cdot \mathbf{X}_j(\omega)\cdot \mathbf{T}_{ji}(\omega)\cdot p_i(\omega') \right.$$

$$\left. \cdot \exp[-i(\omega+\omega')t] \right\rangle_{av} . \tag{3.4}$$

The average over time vanishes unless $\omega + \omega' = 0$. Hence,

$$\Delta E_i = -\tfrac{1}{2}\int\limits_{-\infty}^{+\infty} d\omega\, p_i(\omega)\cdot \mathbf{T}_{ij}(\omega)\cdot \mathbf{X}_j(\omega)\cdot \mathbf{T}_{ji}(\omega)\cdot p_i(-\omega) . \tag{3.5}$$

The dipole interaction tensors $\mathbf{T}_{ij}(\omega)$, $\mathbf{T}_{ji}(\omega)$ can be found from Maxwell's equations. In this chapter we restrict ourselves to using the nonretarded interaction tensor (2.21). The case of retardation is considered in chapter 5.

There remains the question regarding the intensity $p_i(\omega)\,p_i(-\omega)$ of fluctuations of molecule i with frequency ω. The origin of these fluctuations is Planck's radiation field. The probability for molecule i to first emit and than absorb a photon is proportional to the number n_ω of photons with frequency ω. Since the inverse process, i.e. absorption prior to emission, has probability $n_\omega + 1$, we find the average interaction probability to be proportional to

$$\langle n_\omega + \tfrac{1}{2}\rangle = \tfrac{1}{2}\coth(\hbar\omega/2kT) . \tag{3.6}$$

Moreover, the probability that molecule i will absorb and emit a single photon of frequency ω is strongly related to the probability that it will loose energy to the photon field. Each photon-exchange interaction which leads to a final electron state different from the initial one contributes to this effective energy dissipation. Both the energy dissipation and the intensity of fluctuations are proportional to the imaginary part $\mathbf{X}_i''(\omega)$ of the polarizability of molecule i. This is the general statement of the fluctuation-dissipation theorem derived by Callen and Welton in 1951 [65],

$$p_i(\omega)\,p_i(-\omega) = (\hbar/2\pi)\,\mathbf{X}_i''(\omega)\cdot\coth(\hbar\omega/2kT) . \tag{3.7}$$

Inserting Eq. (3.7) into Eq. (3.5) we obtain

$$\varDelta E_i = -(\hbar/4\pi) \int\limits_{-\infty}^{+\infty} d\omega \cdot \coth(\hbar\omega/2kT) \cdot \mathrm{tr}\{\mathbf{X}_i''(\omega)\cdot\mathbf{T}_{ij}\cdot\mathbf{X}_j(\omega)\cdot\mathbf{T}_{ji}\}. \qquad (3.8)$$

Since the real and imaginary parts of the susceptibilities $\mathbf{X}_i(\omega)$, $\mathbf{X}_j(\omega)$ are even and odd functions of frequency, respectively, we may also write

$$\varDelta E_i = -(\hbar/4\pi) \int\limits_{-\infty}^{+\infty} d\omega \cdot \coth(\hbar\omega/2kT) \cdot \mathrm{tr}\{\mathbf{X}_i''(\omega)\cdot\mathbf{T}_{ij}\cdot\mathbf{X}_j'(\omega)\cdot\mathbf{T}_{ji}\}. \qquad (3.9)$$

By adding the energy gain $\varDelta E_j$ caused by fluctuations of molecule j we obtain

$$\varDelta E = -(\hbar/4\pi) \int\limits_{-\infty}^{+\infty} d\omega \cdot \coth(\hbar\omega/2kT) \cdot \mathrm{tr}\{\mathrm{Im}[\mathbf{X}_i(\omega)\cdot\mathbf{T}_{ij}\cdot\mathbf{X}_j(\omega)\cdot\mathbf{T}_{ji}]\} \qquad (3.10)$$

and

$$\varDelta E = -(\hbar/4\pi) \int\limits_{-\infty}^{+\infty} d\omega \cdot \coth(\hbar\omega/2kT) \cdot \mathrm{tr}\{\mathbf{X}_i(\omega)\cdot\mathbf{T}_{ij}\cdot\mathbf{X}_j(\omega)\cdot\mathbf{T}_{ji}\}. \qquad (3.11)$$

In our investigations of the preceding chapter using the oscillator model, it proved convenient to integrate over the frequency along the imaginary axis, where all susceptibilities are real. If we shift the frequency integration in Eq. (3.11) to the imaginary axis as well, we note that the poles of $\mathbf{X}_i(\omega)$, $\mathbf{X}_j(\omega)$ lie in the lower half-plane exclusively. We may close the contour of integration in Eq. (3.11) by adding the vanishing integral along an infinite semicircle in the upper half-plane, as shown in Fig. 7. Shifting this contour to the imaginary axis, we are left with the residues at the poles

$$\hbar\omega_n/2kT = i\pi n \qquad (3.12)$$

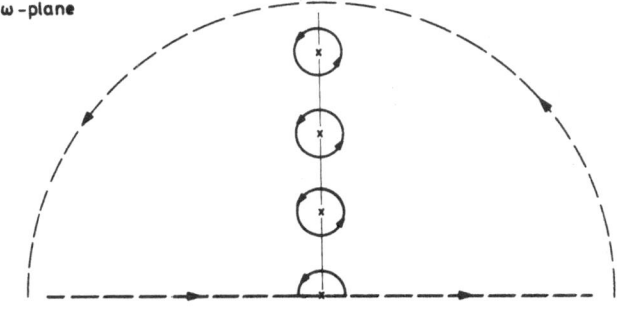

Fig. 7. Contours of integration

of $\coth(\hbar\omega/2kT)$, yielding

$$\Delta E = -kT \sum_{n=0}^{\infty}{}' \ \mathrm{tr}\{\mathbf{X}_i(\omega_n) \cdot \mathbf{T}_{ij} \cdot \mathbf{X}_j(\omega_n) \cdot \mathbf{T}_{ji}\}. \tag{3.13}$$

The prime on the summation symbol in Eq. (3.13) indicates that the term $n=0$ is multiplied by $\frac{1}{2}$, see Fig. 7.

At low temperatures, when the separation between the poles ω_n is infinitely small, we may replace Eq. (3.13) by

$$\Delta E = -(\hbar/2\pi) \int_{0}^{\infty} d\omega \cdot \mathrm{tr}\{\mathbf{X}_i(i\omega) \cdot \mathbf{T}_{ij} \cdot \mathbf{X}_j(i\omega) \cdot \mathbf{T}_{ji}\}. \tag{3.14}$$

This result is equivalent to the findings of Eqs. (2.22) and (2.24) obtained by means of the oscillator model, if the electric susceptibilities $\mathbf{X}_i(\omega)$, $\mathbf{X}_j(\omega)$ of molecules i and j are even functions of $i\omega$, that is, if the poles of $\mathbf{X}_i(\omega)$, $\mathbf{X}_j(\omega)$ lie infinitesimally close to the real frequency axis, as is implied in Eqs. (2.7) and (2.13). This equivalence of the two approaches holds if multiplet interactions are included.

In the alternative case, if the poles of $\mathbf{X}_i(\omega)$, $\mathbf{X}_j(\omega)$ lie at a finite distance from the real frequency axis, we have to reconsider the presented semiclassical approaches and apply quantum electrodynamics. It is certainly a critical procedure simply to add the energy gain resulting from fluctuations of molecules i and j, if molecule j heavily dissipates the energy absorbed by molecule i, and vice versa. Similarly, it is not at all justified to assume that the ground state energy of a damped harmonic oscillator still equals $\frac{1}{2}\hbar\omega$.

3.2. Multiplet Interactions

On applying the above fluctuation approach to multiplets of molecules we note that an induced dipole p_j^{ind} at molecule j not only lowers the energy of the instantaneous dipole p_i^{inst} at molecule i, but also induces further dipoles

$$p_k^{\mathrm{ind}} = p_j^{\mathrm{ind}} \cdot \mathbf{T}_{jk} \cdot \mathbf{X}_k \tag{3.15}$$

at all remaining molecules k. The electric fields $p_k^{\mathrm{ind}} \cdot \mathbf{T}_{ki}$ of these tertiary dipoles k change the energy of dipole i by an amount

$$\Delta_3 E_i = -\frac{1}{2}\sum_{j,k} \langle p_i^{\mathrm{inst}} \cdot \mathbf{T}_{ij} \cdot \mathbf{X}_j \cdot \mathbf{T}_{jk} \cdot \mathbf{X}_k \cdot \mathbf{T}_{ki} \cdot p_i^{\mathrm{inst}} \rangle_{\mathrm{av}}. \tag{3.16}$$

Substituting the Fourier representation (3.3) of p_i^{inst} and applying the fluctuation-dissipation theorem (3.7), we obtain

$$\Delta_3 E_i = -(\hbar/4\pi) \int\limits_{-\infty}^{+\infty} d\omega \coth(\hbar\omega/2kT) \cdot \text{tr}\{\mathbf{X}_i''(\omega) \cdot \mathbf{T}_{ij} \cdot \mathbf{X}_j(\omega)$$
$$\cdot \mathbf{T}_{jk} \cdot \mathbf{X}_k(\omega) \cdot \mathbf{T}_{ki}\} . \tag{3.17}$$

By summing over all molecules i, performing a cyclic permutation of subscripts i, j, k, and applying the general relation

$$\text{Im}(X_i)\,\text{Re}(X_j X_k) + \text{Im}(X_j)\,\text{Re}(X_k X_i) + \text{Im}(X_k)\,\text{Re}(X_i X_j)$$
$$= \text{Im}(X_i X_j X_k) - 2\,\text{Im}(X_i)\,\text{Im}(X_j)\,\text{Im}(X_k) \tag{3.18}$$

we find the total energy gain resulting from triplet interactions to be

$$\Delta_3 E = -(\hbar/4\pi) \int\limits_{-\infty}^{+\infty} d\omega \cdot \coth(\hbar\omega/2kT) \cdot \text{tr}\Big\{\tfrac{1}{3}\sum_{ijk} \text{Im}\,[\mathbf{X}_i \cdot \mathbf{T}_{ij} \cdot \mathbf{X}_j$$
$$\cdot \mathbf{T}_{jk} \cdot \mathbf{X}_k \cdot \mathbf{T}_{ki}] \tag{3.19}$$
$$-\tfrac{2}{3}\sum_{ijk} \mathbf{X}_i'' \cdot \mathbf{T}_{ij} \cdot \mathbf{X}_j'' \cdot \mathbf{T}_{jk} \cdot \mathbf{X}_k'' \cdot \mathbf{T}_{ki}\Big\}.$$

The second term in the braces vanishes if the poles of $\mathbf{X}_i(\omega)$, $\mathbf{X}_j(\omega)$, $\mathbf{X}_k(\omega)$ lie infinitesimally close to the imaginary frequency axis. The first term remains and, by shifting the contour of integration to the imaginary axis analogous to our procedure in the preceding section, becomes

$$\Delta_3 E = -kT \sum_{n=0}^{\infty}{}' \text{tr}\Big\{\tfrac{1}{3}\sum_{ijk} \mathbf{X}_i(\omega_n) \cdot \mathbf{T}_{ij} \cdot \mathbf{X}_j(\omega_n) \cdot \mathbf{T}_{jk} \cdot \mathbf{X}_k(\omega_n)\Big\}. \tag{3.20}$$

Turning to quadruplet interactions, it is obvious from Eq. (3.17) that induction of a further dipole at molecule l yields

$$\Delta_4 E_i = -(\hbar/4\pi) \int\limits_{-\infty}^{+\infty} d\omega \cdot \coth(\hbar\omega/2kT) \,\text{tr}\Big\{\sum_{jkl}{}' \mathbf{X}_i'' \cdot \mathbf{T}_{ij} \cdot \mathbf{X}_j \cdot \mathbf{T}_{jk}$$
$$\cdot \mathbf{X}_k \cdot \mathbf{T}_{kl} \cdot \mathbf{X}_l \cdot \mathbf{T}_{li}\Big\}. \tag{3.21}$$

The summation in Eq. (3.21) includes the possibility that the tertiary dipole p_k^{ind} and the quarternary dipole p_l^{ind} are located at the primary molecule i and the secondary molecule j, respectively. In order to avoid double counting of these contributions, we have to multiply the terms $k = i$, $l = j$ in the sum of Eq. (3.21) by $\tfrac{1}{2}$. This is indicated by the prime on the summation symbol. By summation over all molecules i, and after cyclic permutation of subscripts i, j, k, l, and a rearrangement of the real and imaginary parts of the susceptibilities analogous to Eq. (3.18) we obtain

$$\Delta_4 E = -(\hbar/4\pi) \int_{-\infty}^{+\infty} d\omega \cdot \coth(\hbar\omega/2kT) \, \mathrm{tr}\Big\{ \tfrac{1}{4} \sum_{ijkl} \mathrm{Im}\,[\mathbf{X}_i \cdot \mathbf{T}_{ij} \cdot \mathbf{X}_j$$

$$\cdot \mathbf{T}_{jk} \cdot \mathbf{X}_k \cdot \mathbf{T}_{kl} \cdot \mathbf{X}_l \cdot \mathbf{T}_{li}] \tag{3.22}$$

$$-\tfrac{2}{4} \sum_{ijkl} [\mathbf{X}'_i \cdot \mathbf{T}_{ij} \cdot \mathbf{X}''_j \cdot \mathbf{T}_{jk} \cdot \mathbf{X}''_k \cdot \mathbf{T}_{kl} \cdot \mathbf{X}''_l \cdot \mathbf{T}_{li} + \text{cycl. perm.}]\Big\}.$$

Once more, for the poles of the susceptibilities lying close to the real frequency axis, the second term in the braces of Eq. (3.22) vanishes. Gathering up all pair, triplet and quadruplet contributions, we obtain

$$\Delta E = -kT \sum_{n=0}^{\infty}{}' \, \mathrm{tr}\Big\{ \tfrac{1}{2} \sum_{ij} \mathbf{X}_i(\omega_n) \cdot \mathbf{T}_{ij} \cdot \mathbf{X}_j(\omega_n) \cdot \mathbf{T}_{ji}$$

$$+\tfrac{1}{3} \sum_{ijk} \mathbf{X}_i(\omega_n) \cdot \mathbf{T}_{ij} \cdot \mathbf{X}_j(\omega_n) \cdot \mathbf{T}_{jk} \cdot \mathbf{X}_k(\omega_n) \cdot \mathbf{T}_{ki} \tag{3.23}$$

$$+\tfrac{1}{4} \sum_{ijkl} \mathbf{X}_i(\omega_n) \cdot \mathbf{T}_{ij} \cdot \mathbf{X}_j(\omega_n) \cdot \mathbf{T}_{jk} \cdot \mathbf{X}_k(\omega_n) \cdot \mathbf{T}_{kl} \cdot \mathbf{X}_l(\omega_n) \cdot \mathbf{T}_{li} + \cdots \Big\}.$$

If we assume the poles of $\mathbf{X}_i(\omega)$, $\mathbf{X}_j(\omega),\dots$ to lie at a finite distance from the real frequency axis,

$$\mathbf{X}_i(\omega) = \sum_{k \in i} (2\mathbf{m}_k \omega'_k)^{-1} [(\omega_k - \omega)^{-1} + (\omega_k^* + \omega)^{-1}]; \quad \omega_k = \omega'_k + i\omega''_k$$

$$\mathbf{X}_j(\omega) = \sum_{l \in j} (2\mathbf{m}_l \omega'_l)^{-1} [(\omega_l - \omega)^{-1} + (\omega_l^* + \omega)^{-1}]; \quad \omega_l = \omega'_l + i\omega''_l \tag{3.24}$$

we find the omitted terms in Eqs. (3.19) and (3.21) to be proportional to at least two different distances $\{\omega''_k\}$, $\{\omega''_l\}$. Keeping these terms would certainly be inconsistent with the aspect of independent fluctuations.

3.3. Harmonic Oscillators

We now return to the oscillator model. There are several possible ways of demonstrating its equivalence with the above fluctuation approach for zero distance of the poles of $\mathbf{X}_i(\omega)$, $\mathbf{X}_j(\omega),\dots$ from the real frequency axis. We shall begin with the perturbation approach.

Let us consider an arbitrary ensemble $\{i\}$ of harmonic oscillators with elongation u_i and frequency ω_i. If the force exerted by oscillator i on oscillator j equals $u_i T_{ij}(\omega)$, we may use the Hamiltonian

$$H = \sum_i \tfrac{1}{2} m_i (\dot{u}_i^2 + \omega_i^2 u_i^2) - \tfrac{1}{2} \sum_{ij} u_i T_{ij} u_j. \tag{3.25}$$

Applying the Hamilton formalism, we obtain the equations of motion

$$m_i(\ddot{u}_i + \omega_i^2 u_i) = \sum_{j \neq i} u_j T_{ji} \,. \tag{3.26}$$

In order to find the eigenfrequencies Ω_i of the coupled oscillator system, we put

$$u_i = u_i(\omega) \exp(-i\omega t) \tag{3.27}$$

which yields the secular system

$$m_i(\omega_i^2 - \omega^2) u_i(\omega) - \sum_{j \neq i} u_j(\omega) T_{ji} = 0 \,. \tag{3.28}$$

Assuming for the present that no two unperturbed frequencies ω_i are degenerate, we obtain from perturbation theory (Brillouin-Wigner expansion)

$$\omega^2 = \omega_i^2 + \sum_{j \neq i} \frac{T_{ij} T_{ji}}{m_i m_j (\omega^2 - \omega_j^2)} - \sum_{jk \neq i} \frac{T_{ij} T_{jk} T_{ki}}{m_i m_j m_k (\omega^2 - \omega_j^2)(\omega^2 - \omega_k^2)}$$

$$+ \sum_{jkl \neq i} \frac{T_{ij} T_{jk} T_{kl} T_{li}}{m_i m_j m_k m_l (\omega^2 - \omega_j^2)(\omega^2 - \omega_k^2)(\omega^2 - \omega_l^2)} - + \cdots \,. \tag{3.29}$$

Substituting iteratively the full expression (3.29) for ω^2 in the denominators on the right hand side, we obtain the perturbed eigenfrequencies Ω_i (Rayleigh-Schrödinger expansion)

$$\Omega_i^2 = \omega_i^2 + \sum_{j \neq i} \frac{T_{ij} T_{ji}}{m_i m_j (\omega_i^2 - \omega_j^2)} - \sum_{jk \neq i} \frac{T_{ij} T_{jk} T_{ki}}{m_i m_j m_k (\omega_i^2 - \omega_j^2)(\omega_i^2 - \omega_k^2)}$$

$$+ \sum_{jkl \neq i} \frac{T_{ij} T_{jk} T_{kl} T_{li}}{m_i m_j m_k m_l (\omega_i^2 - \omega_j^2)(\omega_i^2 - \omega_k^2)(\omega_i^2 - \omega_l^2)} \tag{3.30}$$

$$- \sum_{j \neq i} \frac{T_{ij} T_{ji}}{m_i m_j (\omega_i^2 - \omega_j^2)^2} \sum_{k \neq i} \frac{T_{ik} T_{ki}}{m_i m_k (\omega_i^2 - \omega_k^2)} + - \cdots \,.$$

The interaction energy of the oscillator system under investigation is given by the difference between the average quantum energies of the perturbed and the unperturbed modes. Considering first the zero temperature limit we put

$$\Delta E = \sum_i \tfrac{1}{2} \hbar (\Omega_i - \omega_i) \,. \tag{3.31}$$

Taking the square root of Eq. (3.30) and summing over i yields

$$\Delta E = -\tfrac{1}{4}\hbar \sum_i \left\{ \sum_{j\neq i} \frac{T_{ij}T_{ji}}{m_i m_j \omega_i(\omega_j^2 - \omega_i^2)} + \sum_{jk\neq i} \frac{T_{ij}T_{jk}T_{ki}}{m_i m_j m_k \omega_i(\omega_j^2 - \omega_i^2)(\omega_k^2 - \omega_i^2)} \right.$$

$$+ \sum_{jkl\neq i} \frac{T_{ij}T_{jk}T_{kl}T_{li}}{m_i m_j m_k m_l \omega_i(\omega_j^2 - \omega_i^2)(\omega_k^2 - \omega_i^2)(\omega_l^2 - \omega_i^2)} \tag{3.32}$$

$$\left. + \tfrac{1}{2}(\partial/\partial\omega_i^2)\,\omega_i^{-1} \left(\sum_{j\neq i} \frac{T_{ij}T_{ji}}{m_i m_j(\omega_j^2 - \omega_i^2)} \right)^2 + \cdots \right\}.$$

The second order term in Eq. (3.32) agrees with the preliminary expression (2.18) if two oscillators are considered only. By cyclic permutation of subscripts $i, j, k, \ldots$ we obtain

$$\Delta E = -\tfrac{1}{4}\hbar \left\{ \sum_{ij} m_i^{-1} T_{ij} m_j^{-1} T_{ji} \tfrac{1}{2}\Sigma_2 + \sum_{ijk} m_i^{-1} T_{ij} m_j^{-1} T_{jk} m_k^{-1} T_{ki} \tfrac{1}{3}\Sigma_3 \right.$$

$$\left. + \sum_{ijkl} m_i^{-1} T_{ij} m_j^{-1} T_{jk} m_k^{-1} T_{kl} m_l^{-1} T_{li} \tfrac{1}{4}\Sigma_4 + \cdots \right\} \tag{3.33}$$

where

$$\Sigma_2 = 1/\omega_i \omega_j(\omega_i + \omega_j) \tag{3.34}$$

$$\Sigma_3 = (\omega_i + \omega_j + \omega_k)/\omega_i \omega_j \omega_k(\omega_i + \omega_j)(\omega_j + \omega_k)(\omega_k + \omega_i) \tag{3.35}$$

$$\Sigma_4 = \tfrac{1}{3}[(\omega_i + \omega_j + \omega_k + \omega_l)^3 - (\omega_i^3 + \omega_j^3 + \omega_k^3 + \omega_l^3)]/$$

$$\omega_i \omega_j \omega_k \omega_l(\omega_i + \omega_j)(\omega_i + \omega_k)\ldots(\omega_k + \omega_l) \tag{3.36}$$

and generally

$$\Sigma_n(\omega_1, \omega_2, \ldots, \omega_n) = \sum_{i=1}^{n} \omega_i^{-1} \prod_{j=1}^{n}{}' (\omega_j^2 - \omega_i^2)^{-1}. \tag{3.37}$$

The prime on the product sign in Eq. (3.37) indicates that the term $j = i$ is to be excluded.

The explicit representations (3.34)–(3.36) of $\Sigma_n(\omega_i, \omega_j, \ldots)$ show that no poles are left at $\omega_i = \omega_j, \omega_k, \ldots$. We are now free to cancel the restriction to nondegenerate frequencies ω_i. Expression (3.33) is valid for arbitrary frequencies ω_i of the oscillators under investigation. This

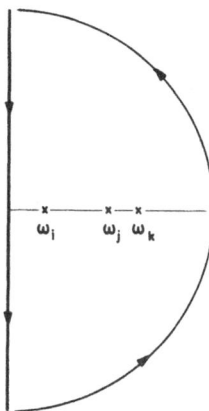

ω-plane

Fig. 8. Complex integration

cancellation of poles arises because two degenerate frequencies strongly repel each other under the action of an external perturbation, whereas the average frequency is maintained.

The two fourth order terms in Eq. (3.32) are combined to form one term in Eq. (3.33), i.e., the second fourth order term in Eq. (3.32) can be shown to compensate for the term $k = i$ omitted in the first fourth order term. A similar fusion of terms also occurs in all higher order contributions.

The representation of the dispersion energy of a system of coupled oscillators by Eqs. (3.33)–(3.36) is the one which is best adapted to practical investigations. From the theoretical point of view it is most appropiate once more to introduce an integration along the imaginary frequency axis. By integrating $\Pi(\omega^2 - \omega_i^2)^{-1}$ along the contour shown in Fig. 8 and shifting this contour to the poles $\omega = \omega_i$ we obtain

$$(\pi i)^{-1} \oint d\omega \prod_{i=1}^{n} (\omega^2 - \omega_i^2)^{-1} = \sum_{i=1}^{n} \omega_i^{-1} \prod_{j=1}^{n}{}' (\omega_i^2 - \omega_j^2)^{-1} . \qquad (3.38)$$

The right hand side of Eq. (3.38) equals $(-1)^{n-1} \Sigma_n$, i.e. by introducing the susceptibilities $\chi_i(\omega)$ of oscillators i according to Eq. (2.7) we have

$$\Sigma_n(\omega_1, \omega_2, \ldots, \omega_n) = (1/\pi i) \int_{-i\infty}^{+i\infty} d\omega \prod_{i=1}^{n} m_i \chi_i(\omega) \qquad (3.39)$$

yielding

$$\Delta E = -(\hbar/4\pi i) \int\limits_{-i\infty}^{+i\infty} d\omega \left\{ \tfrac{1}{2} \sum_{ij} \chi_i T_{ij} \chi_j T_{ji} + \tfrac{1}{3} \sum_{ijk} \chi_i T_{ij} \chi_j T_{jk} \chi_k T_{ki} \right.$$

$$\left. + \tfrac{1}{4} \sum_{ijkl} \chi_i T_{ij} \chi_j T_{jk} \chi_k T_{kl} \chi_l T_{li} + \cdots \right\}. \tag{3.40}$$

By summing over all oscillators attached to the same molecule and over all space directions, we eventually obtain

$$\Delta E = -(\hbar/4\pi i) \int\limits_{-i\infty}^{+i\infty} d\omega \, \mathrm{tr} \left\{ \tfrac{1}{2} \sum_{ij} \mathbf{X}_i \mathbf{T}_{ij} \mathbf{X}_j \mathbf{T}_{ji} + \tfrac{1}{3} \sum_{ijk} \mathbf{X}_i \mathbf{T}_{ij} \mathbf{X}_j \mathbf{T}_{jk} \mathbf{X}_k \mathbf{T}_{ki} \right.$$

$$\left. + \tfrac{1}{4} \sum_{ijkl} \mathbf{X}_i \mathbf{T}_{ij} \mathbf{X}_j \mathbf{T}_{jk} \mathbf{X}_k \mathbf{T}_{kl} \mathbf{X}_l \mathbf{T}_{li} + \cdots \right\}. \tag{3.41}$$

Equation (3.41) agrees with the final expression (3.23) obtained by means of the fluctuation approach if the susceptibilities $\mathbf{X}_i(\omega)$, $\mathbf{X}_j(\omega)$, ... are symmetric with respect to the real frequency axis and if the zero temperature limit is taken.

3.4. State Density Integration

The most elegant and powerful derivation of the interaction energy of a coupled system of oscillators is obtained if the density of the interacting states is expressed by the contour integral over the logarithmic derivative of the respective dispersion function. In order to include finite temperatures, we are now interested in the difference in free energy between the coupled and the uncoupled oscillator system. The free energy of a single harmonic oscillator i equals

$$kT \ln \{2 \sinh(\hbar\omega_i/2kT)\} = \tfrac{1}{2}\hbar \int\limits^{\omega_i} d\omega \coth(\hbar\omega/2kT) \tag{3.42}$$

i.e. we put

$$\Delta E = \sum_i [kT \ln\{2\sinh(\hbar\omega/2kT)\}]_{\omega_i}^{\Omega_i}. \tag{3.43}$$

Let us consider the analytical identity

$$\sum_{\text{zeros}} F(\Omega_n) - \sum_{\text{poles}} F(\omega_n) = (2\pi i)^{-1} \oint d\omega \, F(\omega) \, d\ln G(\omega)/d\omega \tag{3.44}$$

where Ω_n extends over all zeros and ω_n over all poles of $G(\omega)$ within the contour of integration. This contour must not contain poles of $F(\omega)$. If $G(\omega)$ is chosen such that its zeros yield the eigenfrequencies Ω_i of the coupled oscillator system, whereas its poles yield the eigenfrequencies ω_i of the uncoupled oscillator system, we can use Eq. (3.44) directly for summing Eq. (3.43).

This is achieved if

$$F(\omega) = kT \ln \sinh(\hbar\omega/2kT) \tag{3.45}$$

and if $G(\omega)$ equals the ratio of secular determinants derived from Eq. (3.28) for finite and infinite separation. The contour of integration in Eq. (3.44) must enclose the positive real axis. It can be shifted to the imaginary axis, in which case it has to by-pass the branch points of $F(\omega)$ from the right, see Fig. 9. The branch points of $kT \ln \sinh(\hbar\omega/2kT)$ are identical with the poles (3.12) of $\coth(\hbar\omega/2kT)$ according to Eq. (3.42).

Using Eqs. (3.43) to (3.45) we find

$$\Delta E = (kT/2\pi i)\oint d\omega \ln \sinh(\hbar\omega/2kT)\, d\ln G(\omega)/d\omega \tag{3.46}$$

where

$$G(\omega) = \begin{vmatrix} 1 & \dfrac{T_{ji}}{m_i(\omega^2 - \omega_i^2)} & \dfrac{T_{ki}}{m_i(\omega^2 - \omega_i^2)} & \cdots \\[2ex] \dfrac{T_{ij}}{m_j(\omega^2 - \omega_j^2)} & 1 & \dfrac{T_{kj}}{m_j(\omega^2 - \omega_j^2)} & \cdots \\[2ex] \dfrac{T_{ik}}{m_k(\omega^2 - \omega_k^2)} & \dfrac{T_{jk}}{m_k(\omega^2 - \omega_k^2)} & 1 & \cdots \\[2ex] \cdot & \cdot & & \cdots \end{vmatrix} \tag{3.47}$$

according to Eq. (3.28). Expanding $G(\omega)$ with respect to the off-diagonal elements, we find that $\ln G(\omega)$ decreases in proportion to $|\omega|^4$ for large moduli $|\omega|$, i.e., when choosing the right hand contour in Fig. 9 we may omit the integral along the infinite semicircle. After partial integration of the integral along the imaginary axis, we find

$$\Delta E = (\hbar/4\pi i) \int_{-i\infty}^{+i\infty} d\omega \coth(\hbar\omega/2kT) \ln G(\omega) \tag{3.48}$$

with the contour of integration by-passing the poles of $\coth(\hbar\omega/2kT)$ at the right hand side.

We now have to calculate $G(\omega)$ only on the imaginary frequency axis. It is no longer necessary to solve the secular system (3.28) explicitly. $G(\omega)$ is a smoothly varying and rapidly converging function without zeros or poles on the imaginary frequency axis.

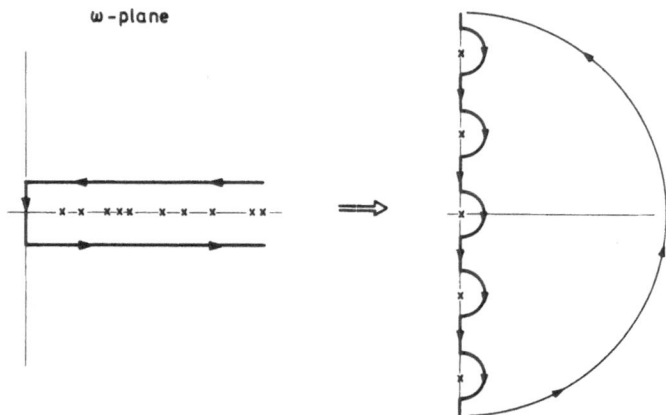

Fig. 9. Contours of integration

By using Eq. (2.7) and joining all oscillators attached to the same molecule and space direction we obtain from Eq. (3.47)

$$G(\omega) = \begin{vmatrix} 1 & -T_{ji}X_i & -T_{ki}X_i & \cdots \\ -T_{ij}X_j & 1 & -T_{kj}X_j & \cdots \\ -T_{ik}X_k & -T_{jk}X_k & 1 & \cdots \\ \cdot & \cdot & \cdot & \cdots \end{vmatrix}. \tag{3.49}$$

Expanding (3.49) with respect to the off-diagonal elements, we again obtain Eqs. (3.23) and (3.41).

The basic trick of the integration method presented here is that of choosing $G(\omega)$ to be the ratio of the dispersion functions of the coupled and the uncoupled oscillator systems. The contour integral around the logarithmic derivative of $G(\omega)$ describes the density of states of the coupled system and, with opposite sign, that of the uncoupled system as well. This choice of $G(\omega)$ guarantees convergence of all relevant integrals.

The application of the analytical identity (3.44) to the van der Waals binding energy in crystals at zero temperature was first reported by Mahan in 1965 [34]. Van Kampen et al. used the same theorem in investigations on the van der Waals energy between half-spaces [35]. The extension to include finite temperatures is due to Ninham et al. [36].

3.5. Screened Interaction Fields

Having derived the general expressions (3.23), (3.41), and (3.48) which cover all multiplet contributions to the dispersion energy, we are able to

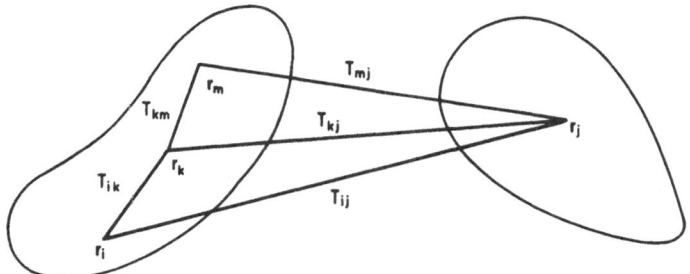

Fig. 10. Summation of pair interactions

reconsider the attraction between macroscopic particles. We now require to find the effect of the multiplet interactions on the findings based on the integration of pair interactions in Sections 2.4–2.7.

There is obviously screening. Let us consider two arbitrary particles 1 and 2, and let us inquire about the external field of a fluctuating dipole p_i at position r_i in particle 1, see Fig. 10. The total field at position r_j outside particle 1 is composed of the direct field $p_i \mathbf{T}_{ij}$, the first-order polarisation field $p_i \mathbf{T}_{ik} \boldsymbol{\chi}_k \mathbf{T}_{kj}$ of all remaining dipoles k in 1, the second-order polarisation field $p_i \mathbf{T}_{ik} \boldsymbol{\chi}_k \mathbf{T}_{km} \boldsymbol{\chi}_m \mathbf{T}_{mj}$ of all remaining dipoles k, m in 1, and so on. We finally obtain the screened interaction tensor

$$\mathbf{T}_{ij}^{\mathrm{scr}} = \mathbf{T}_{ij} + \sum_{k \in 1} \mathbf{T}_{ik} \cdot \boldsymbol{\chi}_k \cdot \mathbf{T}_{kj} + \sum_{km \in 1} \mathbf{T}_{ik} \cdot \boldsymbol{\chi}_k \cdot \mathbf{T}_{km} \cdot \boldsymbol{\chi}_k \cdot \mathbf{T}_{mj} + \cdots . \tag{3.50}$$

Similarly, if position r_j is that of a dipole j in particle 2, we find the screened reaction tensor of dipole j to be

$$\mathbf{T}_{ji}^{\mathrm{scr}} = \mathbf{T}_{ji} + \sum_{l \in 2} \mathbf{T}_{jl} \cdot \boldsymbol{\chi}_l \cdot \mathbf{T}_{li} + \sum_{ln \in 2} \mathbf{T}_{jl} \cdot \boldsymbol{\chi}_l \cdot \mathbf{T}_{ln} \cdot \boldsymbol{\chi}_n \cdot \mathbf{T}_{ni} + \cdots . \tag{3.51}$$

The screened tensors $\mathbf{T}_{ij}^{\mathrm{scr}}$, $\mathbf{T}_{ji}^{\mathrm{scr}}$ account for screening by that particle in which the respective field originates. On substituting $\mathbf{T}_{ij}^{\mathrm{scr}}$, $\mathbf{T}_{ji}^{\mathrm{scr}}$ into Eq. (3.41) and omitting all contributions resulting from interactions within particle 1 or within particle 2 we obtain

$$\Delta E_{12} = -(\hbar/4\pi i) \int_{-i\infty}^{+i\infty} d\omega \, \mathrm{tr} \Bigg\{ \sum_{i \in 1} \sum_{j \in 2} \mathbf{X}_i \mathbf{T}_{ij}^{\mathrm{scr}} \mathbf{X}_j \mathbf{T}_{ji}^{\mathrm{scr}}$$

$$+ \tfrac{1}{2} \sum_{ik \in 1} \sum_{jl \in 2} \mathbf{X}_i \mathbf{T}_{ij}^{\mathrm{scr}} \mathbf{X}_j \mathbf{T}_{jk}^{\mathrm{scr}} \mathbf{X}_k \mathbf{T}_{kl}^{\mathrm{scr}} \mathbf{X}_l \mathbf{T}_{li}^{\mathrm{scr}} \tag{3.52}$$

$$+ \tfrac{1}{3} \sum_{ikm \in 1} \sum_{jln \in 2} \mathbf{X}_i \mathbf{T}_{ij}^{\mathrm{scr}} \mathbf{X}_j \mathbf{T}_{jk}^{\mathrm{scr}} \ldots \mathbf{X}_m \mathbf{T}_{mn}^{\mathrm{scr}} \mathbf{X}_n \mathbf{T}_{ni}^{\mathrm{scr}} + \cdots \Bigg\} .$$

The dispersion energy between two macroscopic particles 1 and 2 can be expressed in terms of the macroscopic screened interaction tensors $\mathbf{T}_{ij}^{scr}$, $\mathbf{T}_{ji}^{scr}$. We do not need the molecular interaction tensors $\mathbf{T}_{ij}$, $\mathbf{T}_{ji}$ any longer.

In order to verify Eq. (3.52) we note that the dispersion energy according to expression (3.41) is derived from an infinite sum over all closed graphs drawn between molecules $i, j, k, \ldots$ as vertices. The vertex i yields the factor $\mathbf{X}_i$, the line connecting i and j yields the factor $\mathbf{T}_{ij}$. The factor $1/n$ in front of the sum of order n in Eq. (3.41) avoids multiple counting of graphs. It can be omitted if at the same time the initial point of each graph is fixed.

The binding energy of particle 1 can be attributed to the graphs which do not leave particle 1, similarly for particle 2. The dispersion energy between particles 1 and 2 arises from the graphs connecting these particles. If we dissect these graphs every time they enter the opposite particle and sum over all subgraphs remaining within the same particle we finally obtain Eq. (3.52). The second order sum in Eq. (3.52) contains the graphs passing once from particle 1 to particle 2 and back. The fourth order sum in Eq. (3.52) contains the graphs passing forth and back twice between particles 1 and 2, and so on.

Expression (3.52) may be further simplified by introducing the macroscopic reaction tensor

$$\mathbf{S}_{ik} = \sum_{j \in 2} \mathbf{T}_{ij}^{scr} \cdot \mathbf{X}_j \cdot \mathbf{T}_{jk}^{scr} \tag{3.53}$$

of particle 2. $\mathbf{S}_{ik}$ is the macroscopic reaction of particle 2 on the field of a screened external dipole at position r_i. Substituting Eq. (3.53) into Eq. (3.52), we obtain

$$\begin{aligned} \Delta E_{12} = -(\hbar/4\pi i) \int_{-i\infty}^{+i\infty} d\omega \operatorname{tr}\Big\{ &\sum_{i \in 1} \mathbf{X}_i \cdot \mathbf{S}_{ii} + \tfrac{1}{2} \sum_{ik \in 1} \mathbf{X}_i \cdot \mathbf{S}_{ik} \cdot \mathbf{X}_k \cdot \mathbf{S}_{ki} \\ &+ \tfrac{1}{3} \sum_{ikm \in 1} \mathbf{X}_i \cdot \mathbf{S}_{ik} \cdot \mathbf{X}_k \cdot \mathbf{S}_{km} \cdot \mathbf{X}_m \cdot \mathbf{S}_{mi} + \cdots \Big\}. \end{aligned} \tag{3.54}$$

All the necessary information about particle 2 needed for calculations of the dispersion energy ΔE_{12} is included in the macroscopic reaction tensor $\mathbf{S}_{ik}$.

3.6. Energy Dissipation

We are now ready to tackle the problem of damping. We have demonstrated the equivalence of the fluctuation approach and the oscillator model in Sections 3.1–3.3, provided that the poles of the molecular susceptibilities $\mathbf{X}_i(\omega)$, $\mathbf{X}_j(\omega)$ lie infinitesimally close to the real frequency axis. How does damping affect these approaches, and is one of them possibly more powerful than the other?

The damping of a single oscillator is caused by its coupling to an infinite ensemble of further particles, which may be photons or other oscillators. The elongation of the oscillator under investigation is no longer periodic, but is given by a superposition of an infinite number of normal modes, making it nearly random. Any systematic elongation caused by an external force fades away, the energy is distributed on all the superimposed normal modes and randomly moves to other particles. On the other hand, the energy is not lost. The normal modes occasionally bring it back, i.e., the oscillator under consideration undergoes random fluctuations. There is a continuous energy exchange between all particles, but never an energy loss.

By ascribing a complex susceptibility to a single oscillator i we admit that we do not have enough information on its coupling to other particles, but describe this coupling globally. We now inquire about the information which we may derive regarding the dispersion energy between two oscillators i and j in spite of their being globally coupled to further particles.

In order to describe damping correctly, we replace each damped oscillator i by an infinite ensemble of undamped oscillators $\{k\}$ in such a manner that the imaginary part of the susceptibility of the ensemble equals that of oscillator i. For real frequencies we find that $\chi_k''(\omega)$ of an undamped oscillator k is a δ-function with an eigenfrequency ω_k minus the same δ-function with $-\omega_k$,

$$\chi_k''(\omega) = (\pi/2m_k\omega_k)\,[\delta(\omega - \omega_k) - \delta(\omega + \omega_k)] \tag{3.55}$$

see Fig. 11. Therefore, we substitute oscillators k whereever $\chi_i''(\omega)$ of the damped oscillator i differs from zero.

$$\chi_i''(\omega) = \sum_k \chi_k''(\omega)\,; \qquad \chi_j''(\omega) = \sum_l \chi_l''(\omega)\,. \tag{3.56}$$

Then, we obtain the real part $\chi_k'(\omega)$ of the susceptibility of oscillator k from the Kramers-Kronig relations

$$\begin{aligned}
\chi_k'(\omega) &= \pi^{-1} \fint_{-\infty}^{+\infty} d\xi\,\chi_k''(\xi)\,(\xi - \omega)^{-1} \\
&= \pi^{-1} \lim_{\varrho \to 0} \left(\int_{-\infty}^{\omega - \varrho} + \int_{\omega + \varrho}^{+\infty} \right) d\xi\,\chi_k''(\xi)\,(\xi - \omega)^{-1},
\end{aligned} \tag{3.57}$$

$$\chi_k''(\omega) = -\pi^{-1} \int_{-\infty}^{+\infty} d\xi\,\chi_k'(\xi)\,(\xi - \omega)^{-1}\,. \tag{3.58}$$

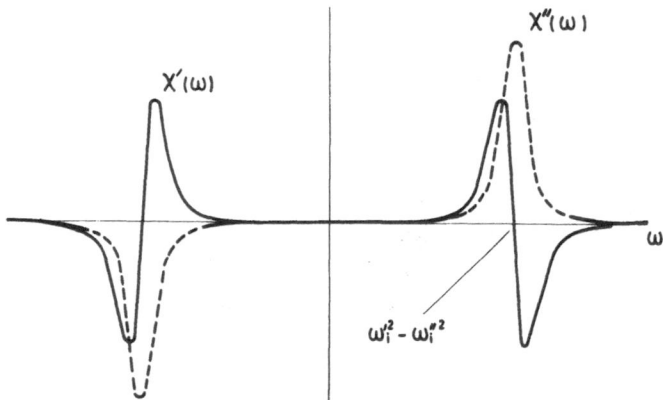

Fig. 11. $\chi(\omega)$ at infinitesimal damping

Let us first apply the fluctuation approach, i.e., let us calculate the dispersion energy between two undamped ensembles $\{k\}$ and $\{l\}$, which replace the damped oscillators i and j in Eq. (3.10). Applying Eqs. (3.55) and (3.57), we find that some precaution is necessary when picking two degenerate oscillators k and l. The integral $\int d\omega \coth(\hbar\omega/2kT)\,\chi_k''(\omega)\,\chi_l'(\omega)$ is not defined in that case. What is the integral over a δ-function times the principle value of an integral, if the branch points of the latter are at the position of the δ-function? These are two conflicting limiting processes. The answer depends critically on the way in which $\chi_k''(\omega)$ approaches the δ-functions (3.55) and $\chi_l'(\omega)$ approaches the principal value (3.57). However, this lack of definition disappears as soon as we add the interchanged integral $\int d\omega \coth(\hbar\omega/2kT)\,\chi_k'(\omega)\,\chi_l''(\omega)$. Defining

$$I_{kl} = \int_{-\infty}^{+\infty} d\omega \coth(\hbar\omega/2kT)\,\{\chi_k''(\omega)\,\chi_l'(\omega) + \chi_k'(\omega)\,\chi_l''(\omega)\} \tag{3.59}$$

we use Eq. (3.57) to obtain

$$\begin{aligned}
I_{kl} = \pi^{-1} \int_{-\infty}^{+\infty} d\omega \coth(\hbar\omega/2kT) \cdot \Bigg\{ &\chi_k''(\omega) \fint_{-\infty}^{+\infty} d\xi\, \chi_l''(\xi)\,(\xi - \omega)^{-1} \\
&+ \chi_l''(\omega) \fint_{-\infty}^{+\infty} d\xi\, \chi_k''(\xi)\,(\xi - \omega)^{-1} \Bigg\}.
\end{aligned} \tag{3.60}$$

Interchanging the order of integration in the second term and the variables ω and ξ we get

$$\begin{aligned}
I_{kl} = -\pi^{-1} \int_{-\infty}^{+\infty} d\omega \fint_{-\infty}^{+\infty} d\xi\, \chi_k''(\omega)\,\chi_l''(\xi) \\
\cdot [\coth(\hbar\omega/2kT) - \coth(\hbar\xi/2kT)]\,(\omega - \xi)^{-1}.
\end{aligned} \tag{3.61}$$

We no longer need to take the principal value of the ξ integral, as it has no pole at $\xi = \omega$. I_{kl} is well-defined, no problems arise when $\chi_k''(\omega)$ and $\chi_l''(\xi)$ approach the δ-functions (3.55). By summing I_{kl} over all oscillators of ensembles k and l and resubstituting Eq. (3.56) we obtain

$$
\begin{aligned}
I_{ij} &= \int_{-\infty}^{+\infty} d\omega \coth(\hbar\omega/2kT) \cdot \operatorname{Im}\left[\chi_i(\omega)\,\chi_j(\omega)\right] \\
&= -\pi^{-1} \int_{-\infty}^{+\infty} d\omega \int_{-\infty}^{+\infty} d\xi\, \chi_i''(\omega)\,\chi_j''(\xi) \\
&\quad \cdot \left[\coth(\hbar\omega/2kT) - \coth(\hbar\xi/2kT)\right] (\omega - \xi)^{-1}.
\end{aligned}
\tag{3.62}
$$

The replacement of the damped oscillators i and j by the undamped ensembles $\{k\}$ and $\{l\}$ finally emerges as the means for finding the physically relevant integral in the presence of damping, i.e. when retarded and advanced susceptibilities have to be used as well. There are good reasons for describing spontaneous and induced fluctuations by advanced and retarded susceptibilities respectively.

In order to avoid further trouble with conflicting limiting processes, we evaluate Eq. (3.62) by assuming

$$
\begin{aligned}
\chi_i(\omega) &= (2m_i\omega_i')^{-1}\left[(\omega_i - \omega)^{-1} + (\omega_i^* + \omega)^{-1}\right]; \\
\chi_j(\omega) &= (2m_j\omega_j')^{-1}\left[(\omega_j - \omega)^{-1} + (\omega_j^* + \omega)^{-1}\right]
\end{aligned}
\tag{3.63}
$$

and using

$$
\chi_i''(\omega) = (2i)^{-1}\left[\chi_i(\omega) - \chi_i(-\omega)\right]; \qquad \chi_j''(\omega) = (2i)^{-1}\left[\chi_j(\omega) - \chi_j(-\omega)\right] \tag{3.64}
$$

for real ω. We close the contours of integration in Eq. (3.62) by adding the respective integrals along infinite semicircles in the upper half-plane and shift this contour to the poles ω_i^*, $-\omega_i$ and ω_j^*, $-\omega_j$ of $\chi_i(-\omega)$ and $\chi_j(-\omega)$. This yields

$$
\begin{aligned}
I_{ij} = &-\pi(2m_i\omega_i')^{-1}(2m_j\omega_j')^{-1} \\
&\cdot \left\{\left[\coth(\hbar\omega_i^*/2kT) - \coth(\hbar\omega_j^*/2kT)\right]/(\omega_i^* - \omega_j^*)\right. \\
&\quad - \left[\coth(\hbar\omega_i/2kT) + \coth(\hbar\omega_j^*/2kT)\right]/(\omega_i + \omega_j^*) \\
&\quad - \left[\coth(\hbar\omega_i^*/2kT) + \coth(\hbar\omega_j/2kT)\right]/(\omega_i^* + \omega_j) \\
&\quad + \left.\left[\coth(\hbar\omega_i/2kT) - \coth(\hbar\omega_j/2kT)\right]/(\omega_i - \omega_j)\right\}.
\end{aligned}
\tag{3.65}
$$

The residues resulting from the poles ω_n of $\coth(\hbar\omega/2kT)$ and $\coth(\hbar\xi/2kT)$ mutually cancel.

By rearranging the right hand side of Eq. (3.65) with respect to the argument of the hyperbolic cotangent and reinterpreting these arguments as resulting from poles in the right-hand half-plane exclusively we obtain

$$
\begin{aligned}
I_{ij} = & -(8im_i\omega_i' m_j\omega_j')^{-1} \oint d\omega \coth(\hbar\omega/2kT) \\
& \cdot \{[(\omega_i-\omega)^{-1}+(\omega_i^*+\omega)^{-1}][(\omega_j-\omega)^{-1}+(\omega_j^*+\omega)^{-1}] \\
& + [(\omega_i^*-\omega)^{-1}+(\omega_i+\omega)^{-1}][(\omega_j^*-\omega)^{-1}+(\omega_j+\omega)^{-1}]\} \,.
\end{aligned}
\tag{3.66}
$$

The second term in the brackets of Eq. (3.66) is the product of the advanced susceptibilities

$$
\chi_i^{\mathrm{adv}}(\omega) = \chi_i^{\mathrm{ret}}(-\omega); \qquad \chi_j^{\mathrm{adv}}(\omega) = \chi_j^{\mathrm{ret}}(-\omega)
\tag{3.67}
$$

which yields

$$
I_{ij} = (2i)^{-1} \int_{-i\infty}^{+i\infty} d\omega \coth(\hbar\omega/2kT)\{\chi_i^{\mathrm{ret}}(\omega)\chi_j^{\mathrm{ret}}(\omega)+\chi_i^{\mathrm{adv}}(\omega)\chi_j^{\mathrm{adv}}(\omega)\} \,.
\tag{3.68}
$$

We obtain an integral along the full imaginary half-axis, which by-passes the poles of $\coth(\hbar\omega/2kT)$ on the right hand side. Compared with our findings in the absence of damping, we find $\chi_i(\omega)\chi_j(\omega)$ replaced by the symmetric expression

$$
\chi_i(\omega)\chi_j(\omega) \Rightarrow \tfrac{1}{2}\{\chi_i^{\mathrm{ret}}(\omega)\chi_j^{\mathrm{ret}}(\omega)+\chi_i^{\mathrm{adv}}(\omega)\chi_j^{\mathrm{adv}}(\omega)\} \,.
\tag{3.69}
$$

By splitting the integral along the imaginary frequency axis into its principal value and the integrals half way around the poles of the hyperbolic cotangent, we find that the principal value vanishes because expression (3.69) is an even function of frequency, whereas the integrals around the poles add up to give

$$
I_{ij} = (2\pi kT/\hbar) \sum_{n=-\infty}^{+\infty} \chi_i(\omega_n)\chi_j(\omega_n)
\tag{3.70}
$$

with ω_n given by Eq. (3.12). Hence

$$
\Delta E = -\tfrac{1}{2}kT \sum_{n=-\infty}^{+\infty} \mathrm{tr}[\mathbf{X}_i(\omega_n)\cdot\mathbf{T}_{ij}\cdot\mathbf{X}_j(\omega_n)\cdot\mathbf{T}_{ji}] \,.
\tag{3.71}
$$

On comparing Eq. (3.71) with the preliminary expression (3.13), we find all poles of $\coth(\hbar\omega/2kT)$ to contribute equally to the dispersion energy. Expression (3.71) yields a dispersion energy larger than that resulting from Eq. (3.13). The correct treatment of damping provides us with a symmetric integral over the susceptibilities involved and increases the dispersion energy relative to the nonsymmetric expression (3.13). This is of particular importance in conductors, where conductivity gives rise to a large imaginary part in the electric susceptibility.

The preceding considerations are based on the fluctuation approach and the understanding that damping arises from a random coupling to an infinite number of further particles. The energy flows randomly to these particles, but occasionally returns. Let us now turn to the oscillator model, which in connection with the state density integration enables the powerful formalism presented in Section 3.4 to be used.

On applying this formalism to damped oscillators, we learn that the secular determinant resulting from Eq. (3.28) is no longer Hermitic. Its eigenfrequencies Ω_i contain a small imaginary component, i.e. the corresponding normal modes decrease exponentially with time. The energy dissipates to further modes which are not considered explicitly. The latter are photons if the oscillators describe electron transitions and electron transitions if the oscillators describe photons.

On the other hand, thermodynamic equilibrium requires an equivalent backflow of energy from the globally described to the explicitly treated modes. We obtain this backflow by using advanced rather than retarded susceptibilities. The true secular system for the eigenfrequencies Ω_i, which properly accounts for the continuous energy exchange between all modes, has to include retarded and advanced susceptibilities in a symmetric manner. The secular determinant then becomes Hermitic and all eigenfrequencies remain strictly real. However, there are several possibilities for constructing such a Hermitic secular determinant. It is difficult to decide on their various merits on a semiclassical basis. The simplest method is to assume that the free energy dissipated from the modes considered explicitly to the modes treated globally equals that dissipated from the latter to the former and to calculate the dispersion energy from

$$\Delta E = \sum_i \mathrm{Re}\left[kT \ln\{2\sinh(\hbar\omega/2kT)\}\right]_{\omega_i}^{\Omega_i}. \qquad (3.72)$$

The logarithm entails that Eq. (3.72) is satisfied if the dispersion function $G(\omega)$ according to Eqs. (3.47), (3.49) is replaced by the symmetric dispersion function

$$G(\omega) \Rightarrow \left[G^{\mathrm{ret}}(\omega)\, G^{\mathrm{adv}}(\omega)\right]^{1/2}. \qquad (3.73)$$

$G(\omega)$ according to Eq. (3.73) is uniquely defined if the branch cuts required by the square root are introduced between corresponding retarded and advanced eigenfrequencies Ω_i and Ω_i^*.

Substituting Eq. (3.73) into Eq. (3.48) and expanding $G(\omega)$ with respect to the off-diagonal elements, we again obtain Eq. (3.68). The symmetric definition of $G(\omega)$ adopted in Eq. (3.73) agrees with the substitution (3.69) derived in the fluctuation approach. By inverting the sign of ω in the term containing $G^{\mathrm{adv}}(\omega)$, we find that the principal value of integral (3.48) again vanishes. We are left with the residues at the poles ω_n of $\coth(\hbar\omega/2kT)$, yielding

$$\Delta E = \tfrac{1}{2}kT \sum_{n=-\infty}^{+\infty} \ln G(\omega_n).$$

(3.74)

It is the integral over $\ln G(\omega)$ along the full imaginary frequency axis, which yields the correct dispersion energy. Any investigation yielding twice the integral along the upper imaginary half-axis accounts incorrectly for the backflow of energy from the modes treated globally to those treated explicitly.

4. Macroscopic Particles

4.1. Homogeneous Isotropic Particles

The calculation of the screened interaction tensors $\mathbf{T}_{ij}^{scr}$ and the dispersion energy ΔE_{12} between two macroscopic particles 1 and 2 is greatly simplified if these particles are homogeneous and isotropic. Isotropy entails that the susceptibilities $\mathbf{X}_i(\omega)$ corresponding to the different positions r_i are scalars which can be separated from the interaction tensors $\mathbf{T}_{ij}$.

$$\mathbf{X}_i \cdot \mathbf{T}_{ij} \cdot \mathbf{X}_j \cdot \mathbf{T}_{jk} \cdot \mathbf{X}_k \cdot \mathbf{T}_{kl} \Rightarrow X_i X_j X_k \cdot \mathbf{T}_{ij} \cdot \mathbf{T}_{jk} \cdot \mathbf{T}_{kl}.$$

(4.1)

Homogeneity enables interaction potentials rather than fields to be considered. Making use of Eq. (2.21) we find

$$\mathbf{T}_{ij} \cdot \mathbf{T}_{jk} \cdot \mathbf{T}_{kl} = V_i V_l \{V_j |r_i - r_j|^{-1} \cdot \mathbf{T}_{jk} \cdot V_k |r_k - r_l|^{-1}\}$$

(4.2)

with the term in the braces representing the reaction potential at position r_l due to a point charge at position r_i.

Moreover, all integrations over particles 1 and 2 can be reduced to integrals over their surface. There are two types of integrals: Those containing the interaction tensors between dipoles located in different particles, and those which also contain interaction tensors between dipoles located in the same particle. The latter integrals are needed for calculating the screened interaction tensors $\mathbf{T}_{ij}^{scr}$, the for   egrals are required for the final integration of ΔE_{12} according to   52). Since in the following sections we shall calculate the screened interaction tensors $\mathbf{T}_{ij}^{scr}$ by strictly macroscopic methods, we sha   ustrate the method of summing microscopic interaction tensors   using the example of interacting spheres.

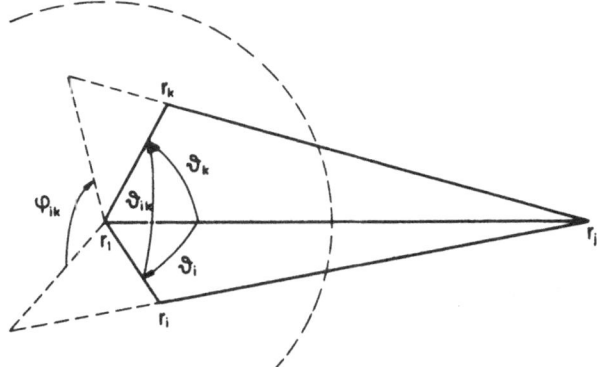

Fig. 12. Polar angles at sphere 1

Let us consider the standard integral

$$\sum_{k\in 1} \mathbf{T}_{ik} \cdot \mathbf{T}_{kj} = V_i V_j \varrho_1 \int_1 d\mathbf{r}_k \, V_k |\mathbf{r}_i - \mathbf{r}_k|^{-1} \cdot V_k |\mathbf{r}_k - \mathbf{r}_j|^{-1} \qquad (4.3)$$

with $\mathbf{r}_i$, $\mathbf{r}_k$ being positions within particle 1 and $\mathbf{r}_j$ lying outside particle 1. By applying

$$V_k |\mathbf{r}_i - \mathbf{r}_k|^{-1} \cdot V_k |\mathbf{r}_k - \mathbf{r}_j|^{-1} = V_k \cdot (|\mathbf{r}_i - \mathbf{r}_k|^{-1} V_k |\mathbf{r}_k - \mathbf{r}_j|^{-1})$$
$$- |\mathbf{r}_i - \mathbf{r}_k|^{-1} \Delta_k |\mathbf{r}_k - \mathbf{r}_j|^{-1} \qquad (4.4)$$

and using the fact that $\Delta_k |\mathbf{r}_k - \mathbf{r}_j|^{-1} = -4\pi \delta(\mathbf{r}_k - \mathbf{r}_j)$ vanishes within particle 1, we obtain

$$\sum_{k\in 1} \mathbf{T}_{ik} \cdot \mathbf{T}_{kj} = V_i V_j \varrho_1 \int_1 d\mathbf{r}_k \, V_k \cdot (|\mathbf{r}_i - \mathbf{r}_k|^{-1} V_k |\mathbf{r}_k - \mathbf{r}_j|^{-1}) . \qquad (4.5)$$

Gauss' integration theorem now yields

$$\sum_{k\in 1} \mathbf{T}_{ik} \cdot \mathbf{T}_{kj} = V_i V_j \varrho_1 \left(\int_{(1)} - \int_{(r_i)} \right) d\mathbf{r}_k \cdot (|\mathbf{r}_i - \mathbf{r}_k|^{-1} V |\mathbf{r}_k - \mathbf{r}_j|^{-1}) . \qquad (4.6)$$

We are left with an integral over the surface of particle 1. However, a small spherical cavity around $\mathbf{r}_i$ has to be excluded, since there is no interaction of dipole at $\mathbf{r}_i$ with itself. All induction fields result by summing over the remaining dipoles. This exclusion of a small spherical cavity is regularly required in microscopic theories on macroscopic susceptibilities.

If particles   a sphere with radius R_i and center $\mathbf{r}_i$ we obtain according to Fig. 12

$$|\mathbf{r}_i - \mathbf{r}_k|^2 = \left. \begin{array}{l} r_1|^2 + |\mathbf{r}_k - \mathbf{r}_1|^2 - 2|\mathbf{r}_i - \mathbf{r}_1| \, |\mathbf{r}_k - \mathbf{r}_1| \cos\vartheta_{ik} \\ = \cos\vartheta_i \cos\vartheta_k + \sin\vartheta_i \sin\vartheta_k \cos\varphi_{ik} \end{array} \right\} \qquad (4.7)$$

$$|\mathbf{r}_k - \mathbf{r}_j|^2 = |\ \ r_1|^2 + |\mathbf{r}_j - \mathbf{r}_1|^2 - 2|\mathbf{r}_k - \mathbf{r}_1| \, |\mathbf{r}_j - \mathbf{r}_1| \cos\vartheta_k \qquad (4.8)$$

where φ_{ik} is the angle between the planes $(r_j - r_1) \times (r_k - r_1)$ and $(r_j - r_1)$ $\times (r_i - r_1)$. Using the generating series of the Legendre polynomials $P_m(\cos\vartheta)$, Eq. (22.9.12) in Ref. [1], we obtain for $|r_j - r_1| > |r_k - r_1|$ $> |r_i - r_1|$

$$|r_i - r_k|^{-1} = \sum_{m=0}^{\infty} [|r_i - r_1|^m / |r_k - r_1|^{m+1}] P_m(\cos\vartheta_{ik}) \tag{4.9}$$

$$|r_k - r_j|^{-1} = \sum_{n=0}^{\infty} [|r_k - r_1|^n / |r_j - r_1|^{n+1}] P_n(\cos\vartheta_k) . \tag{4.10}$$

By substituting Eqs. (4.9) and (4.10) into Eq. (4.6) we obtain

$$\int\limits_{(1)} dr_k \cdot (|r_i - r_k|^{-1} V_k |r_k - r_j|^{-1}) = R_1^2 \int\limits_{-1}^{+1} d\cos\vartheta_k \int\limits_{0}^{2\pi} d\varphi_{ik} \tag{4.11}$$

$$\cdot \sum_{m=0}^{\infty} [|r_i - r_1|^m / R_1^{m+1}] P_m(\cos\vartheta_{ik}) (d/dR_1) \sum_{n=0}^{\infty} [R_1^n / |r_j - r_1|^{n+1}] P_n(\cos\vartheta_k) .$$

The integration over φ_{ik} can be evaluated using the addition theorem for Legendre functions, Eq. (3.11(2)) in Ref. [2],

$$P_m(\cos\vartheta_{ik}) = \sum_{\mu=-m}^{+m} (-1)^\mu P_m^\mu(\cos\vartheta_i) P_m^{-\mu}(\cos\vartheta_k) \exp(i\mu\varphi_{ik}) \tag{4.12}$$

yielding

$$\int\limits_{0}^{2\pi} d\varphi_{ik} P_m(\cos\vartheta_{ik}) = 2\pi P_m(\cos\vartheta_i) P_m(\cos\vartheta_k) . \tag{4.13}$$

Inserting Eq. (4.13) into Eq. (4.11) and using the orthogonality relation for Legendre polynomials, Eq. (22.2.10) in Ref. [1], we obtain

$$\int\limits_{(1)} dr_k \cdot (|r_i - r_k|^{-1} V_k |r_k - r_j|^{-1})$$

$$= 4\pi \sum_{m=1}^{\infty} [m/(2m+1)] [|r_i - r_1|^m / |r_j - r_1|^{m+1}] P_m(\cos\vartheta_i) . \tag{4.14}$$

The reaction potential of a sphere caused by an external point charge is independent of the size of the sphere.

Applying Eq. (4.14) to the integral over a spherical cavity with center r_1 very close to r_i, we are left with the term $m = 1$, yielding

$$V_i \int\limits_{(r_i)} dr_k \cdot (|r_i - r_k|^{-1} V_k |r_k - r_j|^{-1}) = \tfrac{4}{3}\pi V_i |r_i - r_j|^{-1} . \tag{4.15}$$

The decrease in interaction fields caused by the fact that the dipoles do not polarize themselves amounts as usual to $4\pi/3$ of the external field.

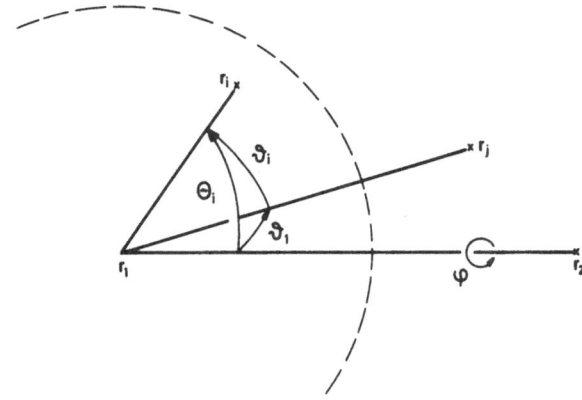

Fig. 13. Rotation of angles

Expanding Eq. (4.15) according to Eq. (4.10) and inserting Eqs. (4.14) and (4.15) into Eq. (4.6) we obtain

$$\sum_{k \in 1} \mathbf{T}_{ik} \cdot \mathbf{T}_{kj} = V_i V_j \tfrac{4}{3} \pi \varrho_1 \sum_{m=1}^{\infty} \frac{m-1}{2m+1} \frac{|\mathbf{r}_i - \mathbf{r}_1|^m}{|\mathbf{r}_j - \mathbf{r}_1|^{m+1}} P_m(\cos \vartheta_i). \qquad (4.16)$$

The multipole expansion of the secondary reaction tensor $\mathbf{T}_{ik}\mathbf{T}_{kj}$ differs from that of the primary reaction tensor $\mathbf{T}_{ij}$ by the additional coefficients $(4\pi/3)\,\varrho_1(m-1)/(2m+1)$. Since each higher order reaction tensor picks up another such coefficient, we obtain

$$\mathbf{T}_{ij}^{\text{scr}} = -V_i V_j \sum_{m=1}^{\infty} \left(1 + \tfrac{4}{3}\pi \varrho_1 X_1 \frac{m-1}{2m+1}\right)^{-1} \frac{|\mathbf{r}_i - \mathbf{r}_1|^m}{|\mathbf{r}_j - \mathbf{r}_1|^{m+1}} P_m(\cos \vartheta_i). \quad (4.17)$$

$\mathbf{T}_{ij}^{\text{scr}}$ is given in terms of its multipole expansion with respect to $\mathbf{r}_i - \mathbf{r}_1$ around the axis $\mathbf{r}_j - \mathbf{r}_1$ directed toward the test position $\mathbf{r}_j$. On rotating angles to a standard direction $\mathbf{r}_2 - \mathbf{r}_1$ according to Fig. 13, we find analogous to Eq. (4.12)

$$P_m(\cos \vartheta_i) = \sum_{\mu=-m}^{+m} (-1)^\mu P_m^\mu(\cos \theta_i) P_m^{-\mu}(\cos \vartheta_1) \exp[i\mu(\varphi_i - \varphi_j)] \qquad (4.18)$$

yielding

$$\mathbf{T}_{ij}^{\text{scr}} = -V_i V_j \sum_{m=1}^{\infty} \left(1 + \tfrac{4}{3}\pi \varrho_1 X_1 \frac{m-1}{2m+1}\right)^{-1} \sum_{\mu=-m}^{+m} (-1)^\mu |\mathbf{r}_i - \mathbf{r}_1|^m$$

$$\cdot P_m^\mu(\cos\theta_i) \exp(i\mu\varphi_i) |\mathbf{r}_j - \mathbf{r}_1|^{-(m+1)} P_m^{-\mu}(\cos\vartheta_1) \exp(-i\mu\varphi_j). \qquad (4.19)$$

We finally obtain a double multipole expansion with respect to $\mathbf{r}_i - \mathbf{r}_1$ and $\mathbf{r}_j - \mathbf{r}_1$ around the standard direction $\mathbf{r}_2 - \mathbf{r}_1$.

4.2. Spheres, Microscopic Approach

Turning to the integration of the macroscopic reaction tensors S_{ik} according to Eq. (3.53) and to the final integration of the dispersion energy ΔE_{12}, we note that we are now dealing with interaction tensors between dipoles located in different particles. Applying Gauss' theorem, we can reduce all integrations over particle 2 to surface integrals without excluding a small cavity. With $U(r_i, r_j)$ and $V(r_j, r_k)$ being arbitrary potential functions we find by generalizing Eqs. (4.3)–(4.6)

$$
\begin{aligned}
&\sum_{j \in 2} \nabla_i \nabla_j U(r_i, r_j) \cdot \nabla_j \nabla_k V(r_j, r_k) \\
&= \nabla_i \nabla_k \varrho_2 \int_{(2)} dr_j \, U(r_i, r_j) \cdot \nabla_j V(r_j, r_k).
\end{aligned}
\tag{4.20}
$$

Let us now assume that particle 2 is a sphere as well, with radius R_2 and center r_2. Then, for the screened interaction tensor T^{scr}_{ik} of a dipole j in 2 with an external dipole k, we find analogous to Eq. (4.19)

$$
\begin{aligned}
T^{scr}_{jk} = &-\nabla_j \nabla_k \sum_{n=1}^{\infty} \left(1 + \tfrac{4}{3}\pi \varrho_2 X_2 \frac{n-1}{2n+1}\right)^{-1} \sum_{v=-n}^{+n} (-1)^v |r_j - r_2|^n \\
&\cdot P_n^v(\cos\theta_j) \exp(i v \varphi_j) |r_k - r_2|^{-(n+1)} P_n^{-v}(\cos\vartheta_2) \exp(-i v \varphi_k)
\end{aligned}
\tag{4.21}
$$

where θ_j, φ_j and ϑ_2, φ_k are the polar angles and azimuths of $r_j - r_2$ and $r_k - r_2$ with respect to the inverted axis $r_1 - r_2$, as shown in Fig. 14.

In order to find $\Sigma T^{scr}_{ij} T^{scr}_{jk}$ we apply Eq. (4.20) to the multipole potentials

$$
U(r_j - r_2) = |r_j - r_2|^n P_n^v(\cos\theta_j) \exp(i v \varphi_j)
\tag{4.22}
$$

$$
V(r_j - r_1) = |r_j - r_1|^{-(m+1)} P_m^\mu(\cos\vartheta_1) \exp(i \mu \varphi_j).
\tag{4.23}
$$

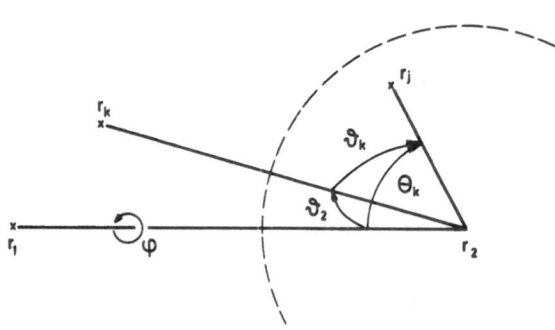

Fig. 14. Rotation of angles

The center of the multipole $V(r_j - r_1)$ can be transposed from r_1 to r_2 by means of the addition theorem

$$|r_j - r_1|^{-(m+1)} P_m^\mu(\cos\vartheta_1) = \sum_{k=\mu}^{\infty} \frac{(m+k)!}{(m-\mu)!(k+\mu)!} \frac{|r_j - r_2|^k P_k^\mu(\cos\theta_j)}{|r_1 - r_2|^{m+k+1}} \quad (4.24)$$

see Fig. 15a. Eq. (4.24) arises from the generating series for Legendre functions (4.9), (4.10), which is recovered for $m = \mu = 0$, by repeated differentiation with respect to angles and separation.

Inserting Eq. (4.24) into Eq. (4.23) we simultaneously change the sign of φ_j in order to keep up with the inverted coordinates at position r_2. By performing the surface integral (4.20), we obtain

$$\sum_{j \in 2} V_j U(r_j - r_2) \cdot V_j V(r_j - r_1) = \frac{4\pi \varrho_2 n}{2n+1} \frac{(m+n)!}{(m-\mu)!(n-\mu)!} \frac{R_2^{2n+1} \delta_{\mu\nu}}{|r_1 - r_2|^{m+n+1}} .$$

$$(4.25)$$

Hence

$$S_{ik} = V_i V_k \sum_{m,n=1}^{\infty} \left(1 + \tfrac{4}{3}\pi \varrho_1 X_1 \frac{m-1}{2m+1}\right)^{-1} \Delta(n,2) [R_2^{2n+1}/|r_1 - r_2|^{m+n+1}]$$

$$\cdot \sum_{\mu} \frac{(m+n)!}{(m+\mu)!(n+\mu)!} \frac{|r_i - r_1|^m}{|r_k - r_2|^{n+1}} P_m^\mu(\cos\theta_1) P_n^\mu(\cos\vartheta_2) \exp[i\mu(\varphi_i + \varphi_k)] \quad (4.26)$$

where

$$\Delta(m,j) = 4\pi \varrho_j X_j m / [(2m+1) + (4\pi/3)\varrho_j X_j(m-1)] . \quad (4.27)$$

We obtain S_{ik} in terms of a double multipole expansion with respect to $r_i - r_1$ around axis $r_2 - r_1$ and with respect to $r_k - r_2$ around the inverted axis $r_1 - r_2$.

The final step for finding the van der Waal energy between spheres 1 and 2 is to integrate the trace of S_{ii}, $\Sigma S_{ik} S_{ki}$, ... over sphere 1. We use

$$\operatorname{tr}\{V_i V_k V(r_i, r_k)\}_{i=k} = \tfrac{1}{2}\Delta_i V(r_i, r_i) \quad (4.28)$$

a)

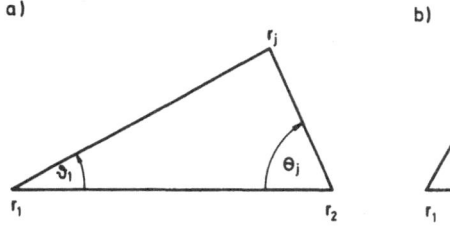
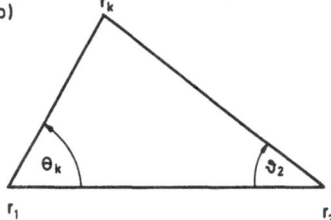

Fig. 15. Addition theorem

where $V(r_i, r_k)$ is an arbitrary potential function, to obtain

$$\sum_{i \in 1} \text{tr}\{V_i V_k V(r_i, r_k)\}_{i=k} = \tfrac{1}{2}\varrho_1 \int_{(1)} dr_i \cdot V_i V(r_i, r_i). \tag{4.29}$$

Then, putting

$$V(r_i, r_k) = [|r_i - r_1|^m / |r_k - r_2|^{n+1}]$$
$$\cdot P_m^\mu(\cos\theta_1) P_n^\mu(\cos\vartheta_2) \exp[i\mu(\varphi_i + \varphi_k)] \tag{4.30}$$

and transposing the multipole potential with respect to r_k from center r_2 to center r_1 by means of the addition theorem (4.24) and Fig. 15b we find

$$\sum_{i \in 1} \text{tr}\{V_i V_k V(r_i, r_k)\}_{i=k}$$
$$= \frac{4\pi\varrho_1 m}{2m+1} \frac{(m+n)!}{(m-\mu)!(n-\mu)!} [R_1^{2m+1} / |r_2 - r_1|^{m+n+1}] \tag{4.31}$$

and

$$\Delta E_{12} = -(\hbar/4\pi) \int_{-\infty}^{+\infty} d\omega \sum_{m,n=1}^{\infty} \binom{2m+2n}{2n}$$
$$\cdot \Delta(m,1)\Delta(n,2)(R_1/r_{21})^{2m+1}(R_2/r_{21})^{2n+1} \tag{4.32}$$

where $r_{21} = |r_2 - r_1|$. In Eq. (4.32) use is made of

$$\sum_\mu \binom{m+n}{m+\mu}\binom{m+n}{n+\mu} = \binom{2m+2n}{2n}. \tag{4.33}$$

We obtain a Taylor expansion of ΔE_{12} with respect to the reduced radii R_1/r_{21}, R_2/r_{21}. If both radii are small compared with the distance r_{21} between the centers, we are left with the dipole-dipole term $m = n = 1$. It is proportional to $R_1^3 R_2^3 / r_{21}^6$, in agreement with our findings based on the integration of pair interactions in Section 2.4. The agreement even includes the frequency integration over the susceptibilities. The dipole susceptibility $\Delta(1, j)$, $j = 1, 2$ of a homogeneous sphere equals $(4\pi/3)\varrho_j$ times the susceptibility of its molecules, see also Eqs. (4.15) and (4.16).

In the opposite case, when the separation $d = r_{21} - R_1 - R_2$ of the spheres is small compared with the radii, we find the reduced radii R_1/r_{21}, R_2/r_{21} to be of the order of one, so that all multipoles make similar contributions to the dispersion energy. We postpone this case to the discussion on the macroscopic approach in the following sections.

Expression (4.32) results from integrating the second order term S_{ii}, i.e., from one reflection of the interaction fields at spheres 1 and 2. In order to integrate the higher order reflection terms $S_{ik}S_{ki}, \ldots$ we repeatedly use the addition theorem (4.24) and integral relation (4.25), yielding

$$\Delta E_{12} = -(\hbar/4\pi) \int_{-\infty}^{+\infty} d\omega \sum_{l=1}^{\infty} l^{-1} \sum_{m_1,\ldots,m_l=1}^{\infty} \sum_{n_1,\ldots,n_l=1}^{\infty} C(m,n)$$

$$\cdot \Delta(m_1, 1)(R_1/r_{21})^{2m_1+1} \Delta(n_1, 2)(R_2/r_{21})^{2n_1+1} \ldots \Delta(n_l, 2)(R_2/r_{21})^{2n_l+1} \qquad (4.34)$$

where

$$C(m,n) = \sum_{\mu} \binom{m_1+n_1}{m_1+\mu} \binom{n_1+m_2}{n_1+\mu} \ldots \binom{m_l+n_l}{m_l+\mu} \binom{n_l+m_1}{n_l+\mu}. \qquad (4.35)$$

We find that each reflection of the interaction fields at sphere 1 and at sphere 2 gives rise to another factor $\Delta(m, 1)(R_1/r_{21})^{2m+1}$ and $\Delta(n, 2)(R_2/r_{21})^{2n+1}$ respectively.

4.3. Spheres, Macroscopic Approach

The clear derivation of the dispersion energy according to Eqs. (4.32), (4.34) from multipole contributions suggests that it is the macroscopic multipole modes rather than the molecular dipole oscillators of spheres 1 and 2 which undergo fluctuations. This becomes even more obvious if we substitute the macroscopic dielectric permeabilities $\varepsilon_1(\omega)$ and $\varepsilon_2(\omega)$ of materials 1 and 2 for the molecular susceptibilities $X_1(\omega)$ and $X_2(\omega)$ according to the law of Clausius-Mosotti

$$\tfrac{4}{3}\pi \varrho_j X_j = (\varepsilon_j - 1)/(\varepsilon_j + 2) \qquad (4.36)$$

yielding

$$\Delta(m, j) = m(\varepsilon_j - 1)/[m\varepsilon_j + (m+1)] . \qquad (4.37)$$

$\Delta(m, j)$ is the 2^m-pole susceptibility of a sphere with dielectric permeability $\varepsilon_j(\omega)$.

Expression (4.34) is built up in a similar manner to expression (3.52), which gives the dispersion energy between particles 1 and 2 in terms of multiplet interactions between their molecules. Multipole m_1 at sphere 1 interacts with multipole n_1 at sphere 2, which in turn interacts with multipole m_2 at sphere 1, and so on. The dispersion energy between spheres 1 and 2 can be obtained by drawing all the closed graphs passing forth and back between the multipoles of spheres 1 and 2 as vertices. Multipole m at sphere j yields the factor $\Delta(m, j)R_j^{2m+1}$, the line passing from multipole m at sphere 1 to multipole n at sphere 2 yields the factor $(m+n)!/(m-\mu)!(n+\mu)!r_{21}^{m+n+1}$. Finally, we have to sum over all rotational wave numbers.

Consequently, it is also possible to represent the dispersion energy in a closed form like in expressions (3.48), (3.49). Let us try to find this representation. What have we really done in the preceding sections?

Ignoring the V_i, V_j operators, which turn the potential of a point charge into the field of a dipole, we performed the following operations: We considered a point charge at position r_i in sphere 1, expanded its potential into multipole potentials at the center r_1 and asked for the resulting external potential $V^{scr}(r_i, r_j)$ of sphere 1 (which corresponds to T_{ij}^{scr}). Then we transformed this potential to sphere 2 by means of the addition theorem (4.24) and asked for the reaction potential $V^{rct}(r_i, r_k)$ of sphere 2 (which corresponds to S_{ik}). This reaction potential lowers the energy of the initiating point charge at r_i, which yields the first order reflection term $l = 1$ to the dispersion energy given by Eq. (4.34), and in turn causes a reaction potential of sphere 1. This alternative transformation and induction of reaction potentials has to be continued until the latter become negligibly small.

Therefore, let us now ask for the exact potential function in the presence of two dielectric spheres. Expanding this potential in terms of the multipoles

$$u_m^\mu(r) = r^m P_m^\mu(\cos\vartheta)\exp(i\mu\varphi)/(m+\mu)! \tag{4.38}$$

$$v_m^\mu(r) = r^{-(m+1)} P_m^\mu(\cos\vartheta)\exp(i\mu\varphi)/(m+\mu)! \tag{4.39}$$

we note that within a single sphere j only $u_m(r)$ is normalizable. By putting

$$V(r - r_j) = b_1 u_m^\mu(r - r_j) \tag{4.40}$$

internally and

$$V(r - r_j) = a_1 u_m^\mu(r - r_j) + a_2 v_m^\mu(r - r_j) \tag{4.41}$$

externally and satisfying the requirement of continuity of $V(r - r_j)$ and of the electric displacement $\varepsilon V V(r - r_j)$ across the surface of sphere j, we obtain

$$\Delta(m, j) R_j^{2m+1} a_1 + a_2 = 0 \tag{4.42}$$

with

$$\Delta(m, j) = m(\varepsilon_j - \varepsilon)/[m\varepsilon_j + (m + 1)\varepsilon] \tag{4.43}$$

and ε being the dielectric permeability externally.

In the presence of two spheres 1 and 2 we construct the external potential from

$$V(r) = \sum_{m=\mu}^{\infty} [a(m, 1)\, v_m^\mu(r - r_1) + a(m, 2)\, v_m^\mu(r - r_2)]. \tag{4.44}$$

Omission of the terms $u_m^\mu(r - r_j)$ guarantees normalizability of $V(r)$ externally as well. Summation over potentials with different rotation wave numbers μ is not necessary if polar coordinates around the axis $r_2 - r_1$ at sphere 1 and around the inverted axis $r_1 - r_2$ at sphere 2 are chosen. By transposing $v_m^\mu(r - r_2)$ from sphere 2 to sphere 1 by means of the addition theorem (4.24) and satisfying the boundary condition (4.42) at the surface of sphere 1, we obtain

$$a(m, 1) + \Delta(m, 1) R_1^{2m+1} \sum_{n=\mu}^{\infty} \frac{(m+n)!}{(n-\mu)!(n+\mu)!} \, a(n, 2) \, r_{21}^{-(m+n+1)} = 0 . \qquad (4.45)$$

Transforming $v_m^\mu(r - r_1)$ from sphere 1 to sphere 2 and satisfying the boundary condition (4.42) at the surface of sphere 2, similarly yields

$$a(n, 2) + \Delta(n, 2) R_2^{2n+1} \sum_{m=\mu}^{\infty} \frac{(n+m)!}{(m-\mu)!(m+\mu)!} \, a(m, 1) r_{21}^{-(m+n+1)} = 0 . \qquad (4.46)$$

We finally obtain the secular system (4.45), (4.46) for the coefficients $a(mj), j = 1, 2$ of the external potential (4.44). Whenever the corresponding secular determinant containing the multipole susceptibilities $\Delta(m, j)$, $j = 1, 2$ has an eigenfrequency Ω_m, there exists a localized potential $V(r)$ of spheres 1 and 2, which may undergo fluctuations. In order to find the dispersion energy between spheres 1 and 2, we apply the integration technique described in Section 3.4. $G(\omega)$ equals the ratio of the secular determinants derived from Eqs. (4.45), (4.46) for finite and infinite separation, with the latter determinant being equal to one.

By summing over all rotational wave numbers μ we obtain

$$\Delta E_{12} = (\hbar/4\pi i) \oint_{-\infty}^{+\infty} d\omega \coth(\hbar\omega/2kT) \sum_{\mu = -\infty}^{+\infty} \ln G(\omega, \mu) \qquad (4.47)$$

where $G(\omega, \mu)$ is the secular determinant given by Eqs. (4.45), (4.46). If we expand these determinants with respect to the off-diagonal elements we recover Eqs. (4.34), (4.35), which are based on the integration of molecular dipole oscillators. The microscopic and the macroscopic methods yield identical results, but the latter procedure is more effective and straigthforward.

The use of electrostatic potentials $V(r)$ between two spheres which fluctuate with frequency Ω_m requires that the separation of the spheres is small enough to neglect retardation. The case of retardation is considered in the next chapter.

4.4. Cylinders

The macroscopic approach to the dispersion energy is based on the interaction of the multipoles of particles 1 and 2, i.e., we have to find exact representations of these multipoles. This is possible in the case of planar, cylindrical, and spherical symmetry. Let us now consider the attraction between two cylinders 1 and 2.

The solution of the Laplace equation in cylindrical coordinates $r = (\varrho, \varphi, z)$ gives the multipoles

$$u_k^m(r) = I_m(k\varrho) \exp(im\varphi) \exp(ikz) \tag{4.48}$$

$$v_k^m(r) = K_m(k\varrho) \exp(im\varphi) \exp(ikz) \tag{4.49}$$

where $I_m(k\varrho)$ and $K_m(k\varrho)$ are modified cylindrical Bessel functions of the first and of the second kind. $I_m(k\varrho)$ vanishes for small arguments, $K_m(k\varrho)$ vanishes for large arguments.

In the presence of a single cylinder j with its axis running through r_j, radius R_j, and dielectric permeability ε_j in a surrounding medium with dielectric permeability ε, we solve the Laplace equation by putting

$$V(r - r_j) = b_1 u_k^m(r - r_j) \tag{4.50}$$

internally and

$$V(r - r_j) = a_1 u_k^m(r - r_j) + a_2 v_k^m(r - r_j) \tag{4.51}$$

externally, see Fig. 16. Continuity of the potential $V(r - r_j)$ and of the electric displacement $-\varepsilon \nabla V(r - r_j)$ across the surface of cylinder j yields

$$\Delta(m, j) a_1 + a_2 = 0 \tag{4.52}$$

where

$$\Delta(m, j) = \frac{(\varepsilon_j - \varepsilon) I_m(kR_j) (\mathrm{d}/\mathrm{d}R_j) I_m(kR_j)}{\varepsilon_j K_m(kR_j) (\mathrm{d}/\mathrm{d}R_j) I_m(kR_j) - \varepsilon I_m(kR_j) (\mathrm{d}/\mathrm{d}R_j) K_m(kR_j)}. \tag{4.53}$$

By using the Wronskian for $I_m(kR_j)$, $K_m(kR_j)$, Eq. (9.6.15) in Ref. [1], we may rewrite Eq. (4.53) to give

$$\Delta(m, j) = \frac{(\varepsilon_j - \varepsilon) R_j (\mathrm{d}/\mathrm{d}R_j) I_m^2(kR_j)}{(\varepsilon_j + \varepsilon) + (\varepsilon_j - \varepsilon) R_j (\mathrm{d}/\mathrm{d}R_j) I_m(kR_j) K_m(kR_j)}. \tag{4.54}$$

In the presence of two cylinders 1 and 2, we construct the external potential symmetrically from Bessel functions of the second kind

$$V(r) = \sum_{m=-\infty}^{+\infty} [a(m, 1) v_k^m(r - r_1) + a(m, 2) v_k^m(r - r_2)]. \tag{4.55}$$

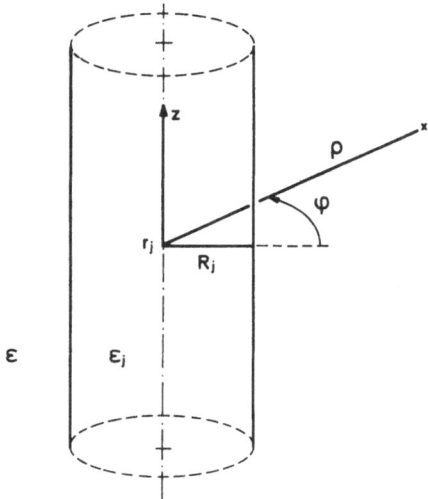

Fig. 16. Cylinder j

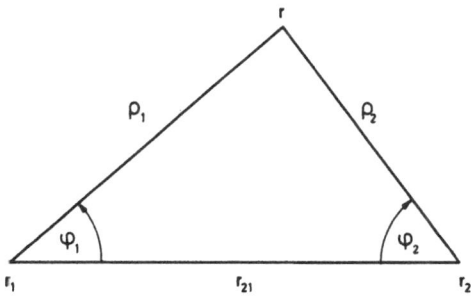

Fig. 17. Inverted cylindrical coordinates

The terms located at cylinder 2 can be transposed to cylinder 1 by means of Graf's addition theorem, Eq. (9.1.79) in Ref. [1]. Using the inverted cylindrical coordinates shown in Fig. 17, we find

$$K_m(k\varrho_2)\exp(im\varphi_2) = \sum_{n=-\infty}^{+\infty} K_{m+n}(k r_{21})\, I_n(k\varrho_1)\exp(in\varphi_1)\,. \qquad (4.56)$$

Applying Eq. (4.56) and satisfying boundary condition (4.52) at the surface of cylinder 1, we obtain

$$a(m,1) + \Delta(m,1)\sum_{n=-\infty}^{+\infty} a(n,2)\, K_{m+n}(k r_{21}) = 0\,. \qquad (4.57)$$

Similarly, satisfaction of the boundary condition (4.52) at the surface of cylinder 2 yields

$$a(n, 2) + \Delta(n, 2) \sum_{m=-\infty}^{+\infty} a(m, 1) K_{m+n}(k r_{21}) = 0 . \qquad (4.58)$$

The secular system (4.57), (4.58) for the coefficients of the localized potential (4.55) can be further reduced by using the fact that the modified cylindrical Bessel functions $I_m(k\varrho)$, $K_m(k\varrho)$ are even functions of order m, yielding

$$a(-m, j) = \pm a(m, j) ; \quad j = 1, 2 . \qquad (4.59)$$

By applying the integration technique described in Section 3.4, we obtain

$$\Delta E_{12} = (\hbar/4\pi i) \oint_{-i\infty}^{+i\infty} d\omega \coth(\hbar\omega/2kT) (L/2\pi) \int_{-\infty}^{+\infty} dk \ln G(\omega, k) \qquad (4.60)$$

where $G(\omega, k)$ is the secular determinant obtained from Eqs. (4.57), (4.58). The expansion of $\ln G(\omega, k)$ with respect to the off-diagonal elements $\Delta(m, j) K_{m+n}(k r_{21})$ gives

$$\ln G(\omega, k) = - \sum_{l=1}^{\infty} l^{-1} \sum_{m_1, \ldots, m_l = -\infty}^{+\infty} \sum_{n_1, \ldots, n_l = -\infty}^{+\infty} \Delta(m_1, 1) K_{m_1+n_1}(k r_{21})$$
$$\cdot \Delta(n_1, 2) K_{n_1+m_2}(k r_{21}) \ldots \cdot \Delta(n_l, 2) K_{n_l+m_1}(k r_{21}) . \qquad (4.61)$$

The dispersion energy between two cylinders 1 and 2 arises in the same way as that between two spheres from repeated reflections of multipole fields from both particles.

By substituting ζ for $k r_{21}$ in Eqs. (4.60), (4.61) we find that ΔE_{12} is proportional to L/r_{21} multiplied by a function of the reduced radii R_1/r_{21}, R_2/r_{21}, with the latter entering ΔE_{12} via the multipole susceptibilities $\Delta(m, j)$, $j = 1, 2$. Using the fact that $K_{m+n}(\zeta)$ decreases more rapidly than $I_m(\zeta R_1/r_{21}) I_n(\zeta R_2/r_{21})$ increases with increasing ζ, we may expand $\Delta(m, j)$ into a Taylor series with respect to $\zeta R_j/r_{21}$. Then, using the integration formula (7.14.2.36) in Ref. [3] we find the main contribution $l = 1$ to ΔE_{12} to be

$$\Delta E_{12} = - (\hbar\langle\omega_1\rangle L/4\pi r_{21}) \sum_{m, n=1}^{\infty} \frac{\Gamma^2(m+n+\frac{1}{2})}{m!(m-1)!n!(n-1)!}$$
$$\cdot (R_1/r_{21})^{2m} (R_2/r_{21})^{2n} \qquad (4.62)$$

where $\langle\omega_1\rangle$ is the first of the characteristic dielectric integrals

$$\langle\omega_l\rangle = i^{-1} \oint_{-i\infty}^{+i\infty} d\omega \coth(\hbar\omega/2kT) \left(\frac{\varepsilon_1 - \varepsilon}{\varepsilon_1 + \varepsilon} \frac{\varepsilon_2 - \varepsilon}{\varepsilon_2 + \varepsilon} \right)^l . \qquad (4.63)$$

At large separations r_{21} we find that the dispersion energy between two cylinders 1 and 2 is proportional to their cross-sections πR_j^2 and obeys a r_{21}^{-5} relationship with regard to separation. This result agrees with our findings based on the integration of pair interactions in Section 2.6. Rather than the frequency integral over the molecular susceptibilities $X_i(\omega)$, $X_j(\omega)$, we now obtain integrals (4.63) over the multipole susceptibilities of the cylinders.

The double hypergeometric series (4.62) is also convergent at small separations $d = r_{21} - R_1 - R_2 \ll R_1$, R_2 of the cylinders. In that case it yields a $d^{-3/2}$ relationship for the dispersion energy with respect to the separation, see Section 4.6.

4.5. Half-Spaces

The solutions of Laplace equation in rectangular coordinates $r = (x, y, z)$ are combinations of plane waves multiplied by exponentially increasing or decreasing functions. In order to treat the case of planar symmetry normal to the x direction, it is convenient to use

$$u_k(r) = \exp(k\,x)\exp[i(k_y y + k_z z)] \tag{4.64}$$

$$v_k(r) = \exp(-k\,x)\exp[i(k_y y + k_z z)] \tag{4.65}$$

where

$$k^2 = k_y^2 + k_z^2 . \tag{4.66}$$

In the presence of a single half-space j extending in the negative x direction, we find that only $u_k(r)$ is normalizable internally. We satisfy the Laplace equation by putting

$$V(r - r_j) = b_1 u_k(r - r_j) \tag{4.67}$$

internally and

$$V(r - r_j) = a_1 u_k(r - r_j) + a_2 v_k(r - r_j) \tag{4.68}$$

externally. It is convenient to choose r_j to be the intersection of the surface of half-space j with the x axis, $r_j = (x_j, 0, 0)$. Continuity of $V(r - r_j)$ and of the electric displacement across the surface yields

$$\Delta(j)a_1 + a_2 = 0 \tag{4.69}$$

where

$$\Delta(j) = (\varepsilon_j - \varepsilon)/(\varepsilon_j + \varepsilon) \tag{4.70}$$

and ε_j and ε are the dielectric permeabilities internally and externally, respectively.

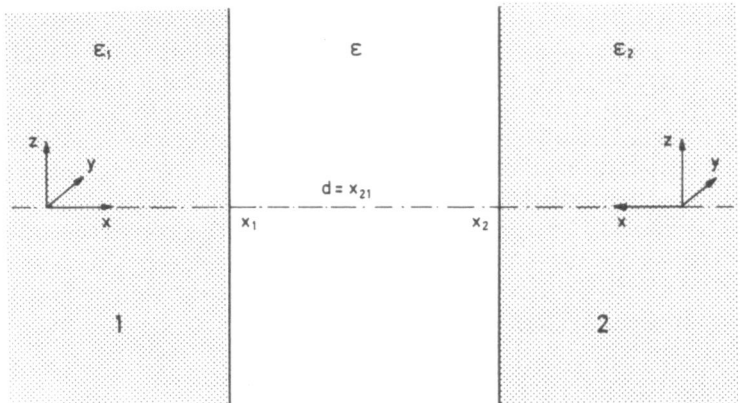

Fig. 18. Attracting half-spaces

We now consider two half-spaces 1 and 2 opposing each other at distance $d = x_2 - x_1$, see Fig. 18. Since the second half-space extends in the positive x direction, we have to interchange $u_k(r - r_j)$ and $v_k(r - r_j)$ in order to satisfy the Laplace equation across its surface. However, in order to continue with the procedure used in the presence of spheres or cylinders, we shall invert the x direction. Hence

$$V(r) = a(1)\, v_k(r - r_1) + a(2)\, v_{-k}(r - r_2) \,. \tag{4.71}$$

Transposing the external potential $v_{-k}(r - r_2)$ from half-space 2 to half-space 1 and satisfying boundary condition (4.69) across its surface, we find

$$a(1) + \Delta(1)\, a(2) \exp[-k(x_2 - x_1)] = 0 \,. \tag{4.72}$$

Similarly,

$$a(2) + \Delta(2)\, a(1) \exp[-k(x_2 - x_1)] = 0 \,. \tag{4.73}$$

We obtain a secular system of order two, i.e., the resulting ratio of secular determinants for finite and infinite separation d is

$$G(\omega, k) = 1 - \Delta(1)\, \Delta(2) \exp(-2kd) \,. \tag{4.74}$$

Application of the integration technique described in Section 3.4 yields the dispersion energy

$$\Delta E_{12} = (\hbar/4\pi i) \oint_{-i\infty}^{+i\infty} d\omega \coth(\hbar\omega/2kT)(L/2\pi)^2 \int_{-\infty}^{+\infty} dk_y \int_{-\infty}^{+\infty} dk_z \ln G(\omega, k) \,. \tag{4.75}$$

The integrals over k_y, k_z can be converted into an integral over k according to $\int dk_y \int dk_z = 2\pi \int k \, dk$. Then, substituting ζ for kd, we obtain

$$\Delta E_{12} = (\hbar/4\pi i) \int_{-i\infty}^{+i\infty} d\omega \coth(\hbar\omega/2kT) (L^2/2\pi d^2)$$

$$\int_0^\infty d\zeta\,\zeta \ln(1 - \Delta(1)\Delta(2)e^{-2\zeta}) \,. \tag{4.76}$$

Expansion of the logarithm and integration over ζ yields

$$\Delta E_{12} = -(\hbar L^2/32\pi^2 d^2) \sum_{l=1}^\infty \langle\omega_l\rangle/l^3 \tag{4.77}$$

where $\langle\omega_l\rangle$ is given by (4.63).

The dispersion energy between two half-spaces is proportional to the inverse square of their separation d. Expression (4.76) was first derived by Lifshitz in 1955 on the basis of the fluctuation approach [28]. The macroscopic investigations presented here were reported by van Kampen et al. in 1968 [35].

We stress the fact that the frequency integral over the electric susceptibilities of the particles involved has to be extended along the full imaginary axis in order to cover the case of damping correctly.

4.6. Small Separations

We now have treated the attraction between spheres, cylinders and half-spaces by equivalent procedures. We have solved the Laplace equation in spherical, cylindrical, or rectangular coordinates, required normalizability of the potentials within these particles and allowed for non-normalizable potentials externally only in so far as these potentials result from transposing a normalizable potential located at a different particle. We obtained localized solutions of the Laplace equation which may fluctuate under the action of an external perturbance (photons).

The dispersion energy is generally found to be a function of the reduced radii and diameters R_1/r_{21}, R_2/r_{21}. At large separations we find the dispersion energy between spheres to be proportional to their volumes $(4\pi/3)(R_1/r_{21})^3$ and $(4\pi/3)(R_2/r_{21})^3$ and, consequently, to vary with r_{21}^{-6}; that between cylinders is proportional to their reduced length L/r_{21} and to their reduced cross-sections $\pi(R_1/r_{21})^2$ and $\pi(R_2/r_{21})^2$ and, consequently, varies with r_{21}^{-5}. If higher order reflection terms $l \geq 2$ are considered, we find that each additional reflection gives rise to another reduced volume or reduced cross-section, so that a r_{21}^{-6l} and a $r_{21}^{-(4l+1)}$

relationship between dispersion energy and separation results in the case of spheres and cylinders, respectively.

A distinction between large and small separations for half-spaces is no longer relevant. The separation d cannot exeed their diameters, i.e., there is merely the case of small separations. We find a $(L/d)^2$ relationship between dispersion energy and separation, which holds inclusive of the higher order reflection terms $l \geq 2$. We now ask whether this general result could give us an indication on the respective behavior of spheres and cylinders at small separations.

The reason why all multiple reflection terms show the same dependence on the separation as does the first order reflection term, is the fact that the dispersion function $G(\omega, k)$ between half-spaces, by not containing any diameter, does not contain any reduced diameter. On the other hand, we cannot expect the dispersion functions $G(\omega, \mu)$ between spheres or cylinders to depend strongly on the reduced radii R_j/r_{21} if the separation of these particles is so small that the rear boundaries are hardly noticed by the interacting multipoles. If the dispersion functions should become independent of R_j/r_{21} at small separations d, we may conclude that the relationships between dispersion energy and separation found on the basis of the first order reflection terms given by Eqs. (4.32) and (4.62) are reproduced by the multiple reflection terms $l \geq 2$.

Let us therefore try to find approximate dispersion functions which are valid at small separations of the interacting particles. The dispersion function between half-spaces is split up depending on the translational behavior of the interacting waves parallel to the surfaces. Each mode $v_k(r - r_1)$ localized at half-space 1 interacts with only one mode $v_{-k}(r - r_2)$ localized at half-space 2. Turning to parallel cylinders 1 and 2 we note that the dispersion function is still separable according to the translational behavior of the modes $v_k^m(r - r_1)$ and $v_k^m(r - r_2)$ parallel to the cylinder axes. Only those modes which are characterized by different angular wave numbers m interact. Nevertheless, we expect the effective wave numbers of interacting modes to be similar in the case of small separations (see Fig. 19) suggesting that

$$m/R_1 \simeq n/R_2 \, . \tag{4.78}$$

Applying second order perturbation theory to the secular system (4.57), (4.58) for the coefficients $a(m, j)$ we find the approximate dispersion relation

$$1 - \Delta(m, 1) \sum_{n=-\infty}^{+\infty} \Delta(n, 2) \, K_{m+n}^2(k r_{21}) = 0 \, . \tag{4.79}$$

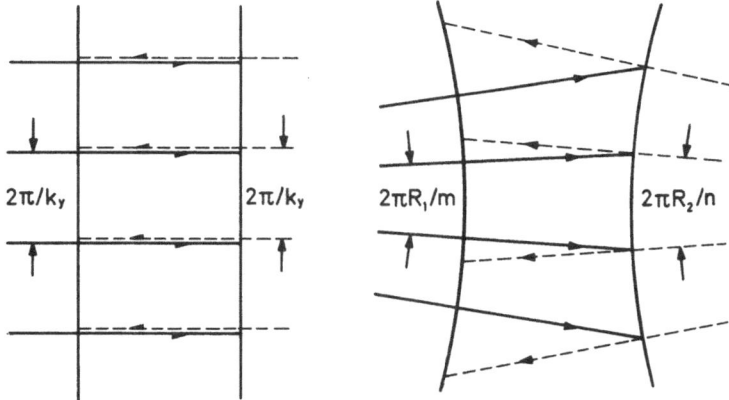

Fig. 19. Reflection from half-spaces and cylinders

Substituting $\Delta(m, 1)$ and $\Delta(n, 2)$ according to (4.54) and neglecting the second terms in the denominators we obtain

$$1 - \Delta(1)\,\Delta(2)\, R_1(\mathrm{d}/\mathrm{d}R_1)\, R_2(\mathrm{d}/\mathrm{d}R_2) \sum_{n=-\infty}^{+\infty} I_m^2(kR_1)$$
$$\cdot K_{m+n}^2(k r_{21})\, I_n^2(k R_2) = 0\,. \tag{4.80}$$

Application of the asymtotic expansions (9.7.1) and (9.7.2) in Ref. [1] yields

$$I_m(kR_1)\,K_{m+n}(k r_{21})\,I_n(kR_2) \simeq (8\pi k^3 R_1 R_2 r_{21})^{1/2}\, \exp\left[-k(r_{21} - R_1 - R_2)\right]$$
$$\left(1 - \frac{(2m+1)(2m-1)}{8kR_1}\right)\left(1 + \frac{(2m+2n+1)(2m+2n-1)}{8k r_{21}}\right) \tag{4.81}$$
$$\cdot\left(1 - \frac{(2n+1)(2n-1)}{8kR_2}\right).$$

The right hand side of Eq. (4.81) shows a maximum in its dependence on n at $(m+n)/r_{21} = n/R_2$, which verifies the suggested strong interaction of modes with similar angular wave number yielding Eq. (4.78). The exponential in Eq. (4.81) indicates that the interaction of modes decreases with $\exp(-2kd)$, as is true in the case of half-spaces.

The summation over n in Eq. (4.81) can be performed explicitly by applying the addition theorem (4.56). Using

$$\sum_{n=-\infty}^{+\infty} K_{m+n}^2(k r_{21})\, I_n^2(kR_2) = (2\pi)^{-1} \int_0^{2\pi} \mathrm{d}\varphi\, K_m^2(k\varrho) \tag{4.82}$$

where

$$\varrho^2 = r_{21}^2 + R_2^2 - 2r_{21}R_2 \cos\varphi \tag{4.83}$$

we find

$$1 - \Delta(1)\Delta(2)R_1(\mathrm{d}/\mathrm{d}R_1)R_2(\mathrm{d}/\mathrm{d}R_2)I_m^2(kR_1)(2\pi)^{-1}\int_0^{2\pi}\mathrm{d}\varphi\, K_m^2(k\varrho)=0. \tag{4.84}$$

Dispersion relation (4.84) yields the frequency shift of the modes localized at cylinder 1 relative to the case of infinite separation. It also covers the frequency shift of the modes localized at cylinder 2 which match those at cylinder 1 according to Eq. (4.78). There is an exponential decrease in frequency shifts with increasing separation $d = r_{21} - R_1 - R_2$, but only a smooth dependence on the radii R_j. Hence, we conclude that the multiple reflection terms exhibit the same dependence on the separation d as do the first order reflection terms. By substituting Eq. (4.84) for $G(\omega, k)$ in Eq. (4.60) and summing over all angular wave numbers m, we obtain

$$\Delta E_{12} = -(\hbar\langle\omega_1\rangle L/4\pi)\,R_1(\mathrm{d}/\mathrm{d}R_1)\,R_2(\mathrm{d}/\mathrm{d}R_2)\,(2\pi)^{-2}\int_{-\infty}^{+\infty}\mathrm{d}k$$
$$\sum_{m=-\infty}^{+\infty}I_m^2(kR_1)\int_0^{2\pi}\mathrm{d}\varphi\, K_m^2(k\varrho)\,. \tag{4.85}$$

Similar to Eq. (4.82) we have

$$\sum_{m=-\infty}^{+\infty}I_m^2(kR_1)\,K_m^2(k\varrho)=(2\pi)^{-1}\int_0^{2\pi}\mathrm{d}\psi\, K_0^2(k\sigma) \tag{4.86}$$

where

$$\sigma^2 = \varrho^2 + R_1^2 - 2\varrho R_1 \cos\psi\,. \tag{4.87}$$

Integrating over k in Eq. (4.85) we obtain

$$\Delta E_{12} = -\tfrac{1}{16}\hbar\langle\omega_1\rangle LR_1(\mathrm{d}/\mathrm{d}R_1)R_2(\mathrm{d}/\mathrm{d}R_2)(2\pi)^{-2}\int_0^{2\pi}\mathrm{d}\varphi\int_0^{2\pi}\mathrm{d}\psi\,\sigma^{-1}\,. \tag{4.88}$$

Using second order perturbation theory, we find that Eq. (4.88) agrees with Eqs. (4.60), (4.61) for $l = 1$. Expression (4.62) is recovered by expanding σ with respect to R_j/r_{21}, which requires large separations, $R_j/r_{21} \ll 1$. At small separations $d \ll R_j$, it is convenient to represent the integral in Eq. (4.88) by the complete elliptic integral $K(z)$ of the first kind, $z = 4\varrho R_1/(\varrho + R_1)^2$, to approximate $K(z)$ at $z \simeq 1$ according to Eq. (17.3.26) in Ref. 1, and then to perform the φ integral. We finally obtain the main term $l = 1$ of the series

$$\Delta E_{12} = -(\hbar L/16\pi)\,(R_1 R_2/(2d)^3(R_1 + R_2))^{1/2}\sum_{l=1}^{\infty}\langle\omega_l\rangle/l^3\,. \tag{4.89}$$

The multiple reflection terms $l \geq 2$ in Eq. (4.89) are added in accordance with our findings on the selective interaction of modes showing similar angular wave number, yielding dispersion relation (4.84). An explicit proof of Eq. (4.89) for arbitrary orders l may be based on the fact that reflection fields which cyclicly reproduce themselves after l reflections can be represented in terms of fields which gain a phase factor $\exp(-2\pi i n/l)$ at each reflection (Floquet's theorem). This was reported by Langbein in 1971 [95, 99].

We again obtain the $d^{-3/2}$ dependence of the dispersion energy on the separation which was found in Section 2.5 by integrating pair interactions, with the integral over the molecular susceptibilities being replaced by that over the multipole susceptibilities of the interacting particles.

The reasoning used for finding the dispersion energy between cylinders at small separations is applicable in a similar manner to the dispersion energy between spheres. The mathematical effort is even less in the latter case. We may, for instance, sum the main term in Eq. (4.32) explicitly by substituting the limiting value $\Delta(m, j) \to \Delta(j)$ for large m, yielding

$$\Delta E_{12} = -(\hbar \langle \omega_1 \rangle/16\pi)(R_1 R_2/r_{21}) \sum_{\mu, \nu = 1}^{4} (-1)^{\mu + \nu}$$

$$\cdot [r_{21} - R_1 \sin(\mu\pi/2) - R_2 \sin(\nu\pi/2)]^{-1} . \qquad (4.90)$$

The term $\mu = \nu = 1$ in Eq. (4.90) yields a d^{-1} dependence of the dispersion energy on the separation d.

Summarizing the findings on spheres, cylinders, and half-spaces we obtain

$$\Delta E_{12} = -(\hbar/32\pi^2)[2\pi R_1 R_2/(R_1 + R_2)]^{1 - n/2}$$

$$\cdot \Gamma(1 + \tfrac{1}{2}n) L^n d^{-(1 + \frac{1}{2}n)} \sum_{l = 1}^{\infty} \langle \omega_l \rangle/l^3 \qquad (4.91)$$

with $n = 0$ for spheres, $n = 1$ for cylinders and $n = 2$ for half-spaces in accord with Eq. (2.49).

4.7. Multilayers

The strength of the above macroscopic procedure lies in the separate calculation of the multipole susceptibilities of the single particles and the total external potential and energy. If the multipole susceptibilities $\Delta(m, j)$ are completely defined, we no longer require the detailed composition of the interaction partners. We can thus extend the theory to the attraction between all particles, which make the effort involved in calculating the multipole susceptibilities reasonable. This applies to

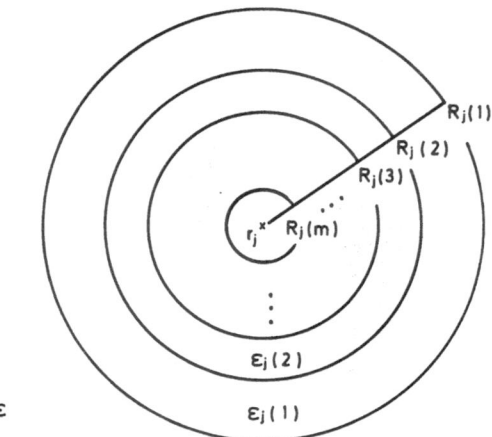

Fig. 20. Concentric layers

spheres and cylinders whose surface is covered with adsorbate layers and to the dispersion energy in multilayer structures.

Let us consider a sphere or cylinder j composed of m concentric layers having different dielectric permeabilities $\varepsilon_j(n), n = 1, \ldots m$, as shown in Fig. 20. In order to satisfy the Laplace equation in the presence of this particle, we construct $V(r - r_j)$ from the normalizable potentials $u(r - r_j)$ according to Eqs. (4.38) or (4.48) in the innermost layer, where $|r - r_j| < R_j(m)$, and from linear combinations of the two potential functions $u(r - r_j)$ and $v(r - r_j)$ according to Eqs. (4.38), (4.39) or (4.48), (4.49) in all intermediate layers where $1 < n < m$, and externally. The requirement of continuity of the potential and the electric displacement across all interfaces $R_j(n), n = 1, \ldots, m$ yields $2m$ equations for the $2m + 1$ coefficients of the potential, i.e. by eliminating all internal coefficients, there remains one relationship between the external coefficients a_1 and a_2 analogous to the expressions (4.42) or (4.52). $\Delta(m, j)$ is given by the ratio of two determinants and depends drastically on the $2m$-th power of the ratio $R_j(n + 1)/R_j(n)$ of successive radii: If m is large, so that $[R_j(n + 1)/R_j(n)]^{2m} \ll 1$, there is no longer an effect from the inner layers. Therefore, the question as to whether the inner layers contribute to ΔE_{12} or not, depends on the number of multipoles m needed for integrating the dispersion energy. If the separation of the particles under consideration is large, we need only a few terms m, n for convergence of Eqs. (4.32) and (4.62), which means that all layers contribute to ΔE_{12}. On the other hand, if the separation of the particles under consideration is small, so that expressions (4.32) and (4.62) converge only slowly, only the effect of the outer layers remains. This general prediction of the theory

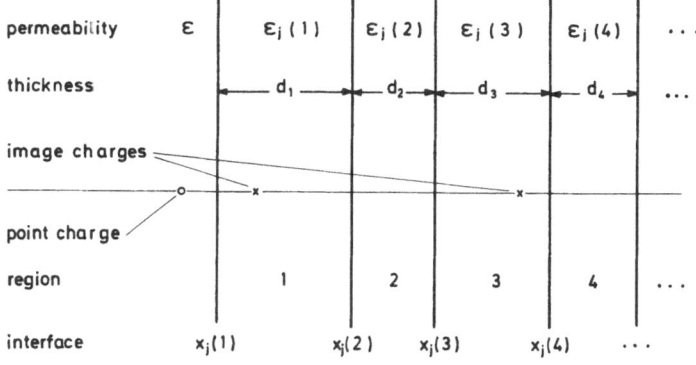

Fig. 21. Infinite layer system

has been verified experimentally by Tabor and Israelachvili for the example of stearic acid films on mica [152].

A most obvious representation of the dispersion energy between half-spaces and multilayers arises by combining the microscopic and the macroscopic approaches. Let us consider microscopic fluctuations and macroscopic reaction fields. It is well-known from electrostatics that the reaction potential of a half-space j caused by an external point charge q equals that of an image charge $q(\varepsilon_j - \varepsilon)/(\varepsilon_j + \varepsilon)$ in the interior, whereas the potential within half-space j equals that of an external point charge $2q\varepsilon_j/(\varepsilon_j + \varepsilon)$. If, rather than a half-space, we consider the infinite layer system shown in Fig. 21, we have to satisfy additional continuity conditions for the potential at all interfaces $x_j(n)$ between regions $n-1$ and n. In addition to the image charge $q(\varepsilon_j(1) - \varepsilon)/(\varepsilon_j(1) + \varepsilon)$ with respect to the surface $x_j(1)$, we need an image charge $q(\varepsilon_j(2) - \varepsilon_j(1))/(\varepsilon_j(2) + \varepsilon_j(1))$ with respect to the interface $x_j(2)$. The potential of the latter, in turn, violates the continuity conditions for the potential with respect to the surface $x_j(1)$, therefore we have to introduce another external image charge. In order to construct the reaction potential of the multilayer system under consideration in general order we obtain the following scheme:

– The continuity conditions for the potential and the electric displacement at the interface $x_j(n)$ require the addition of a reflection potential with a weight factor $(\varepsilon_j(n-1) - \varepsilon_j(n))/(\varepsilon_j(n-1) + \varepsilon_j(n))$ in region $n-1$ and transmission of the potential with a weight factor $2\varepsilon_j(n)/(\varepsilon_j(n-1) + \varepsilon_j(n))$ to region n.

– Reflections and transmissions of the potential of this type are conveniently represented by the graphs shown in Fig. 22, with the respective weight factor given on the right.

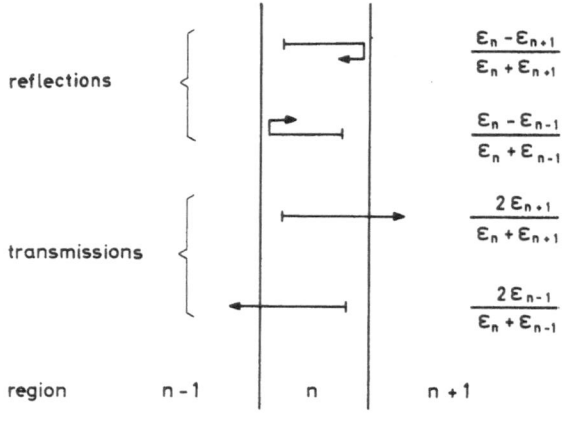

Fig. 22. Graphs

– We find all contributions to the reaction potential of multilayer system j by constructing all diagrams of graphs which start and end externally. The weight factor of a diagram is the product of the weight factors of its graphs. The distance of the corresponding image charge is the sum over the graph lengths.

Knowing the reaction potentials, we can calculate the dispersion energy. We find the dispersion energy ΔE_{12} between two multilayers 1 and 2 by connecting the diagrams representing the reaction potential of multilayer 1 with the diagrams representing that of multilayer 2. The weight factor of each closed diagram is the product of weight factors of its graphs. The effective distance entering ΔE_{12} equals half the sum over the graph lengths. Diagrams consisting of l equivalent paths receive the additional weight factor $1/l$ in order to avoid multiple counting. The main contributions to ΔE_{12} according to this scheme are shown in Fig. 23.

We obtain

$$
\begin{aligned}
\Delta E_{12} = -(\hbar/32\pi^2) \int_{-\infty}^{+\infty} d\omega \Bigg\{ &\frac{\varepsilon - \varepsilon_1(1)}{\varepsilon + \varepsilon_1(1)} \frac{\varepsilon - \varepsilon_2(1)}{\varepsilon + \varepsilon_2(1)} [x_2(1) - x_1(1)]^{-2} \\
+ &\frac{\varepsilon - \varepsilon_1(1)}{\varepsilon + \varepsilon_1(1)} \frac{2\varepsilon\varepsilon_2(1)}{(\varepsilon + \varepsilon_2(1))^2} \frac{\varepsilon_2(1) - \varepsilon_2(2)}{\varepsilon_2(1) + \varepsilon_2(2)} [x_2(2) - x_1(1)]^{-2} \\
+ &\frac{\varepsilon - \varepsilon_2(1)}{\varepsilon + \varepsilon_2(1)} \frac{2\varepsilon\varepsilon_1(1)}{(\varepsilon + \varepsilon_1(1))^2} \frac{\varepsilon_1(1) - \varepsilon_1(2)}{\varepsilon_1(1) + \varepsilon_1(2)} [x_2(1) - x_1(2)]^{-2} \\
+ &\tfrac{1}{2} \left(\frac{\varepsilon - \varepsilon_1(1)}{\varepsilon + \varepsilon_1(1)}\right)^2 \left(\frac{\varepsilon - \varepsilon_2(1)}{\varepsilon + \varepsilon_2(1)}\right)^2 [2x_2(1) - 2x_1(1)]^{-2} + \cdots \Bigg\}
\end{aligned}
\tag{4.92}
$$

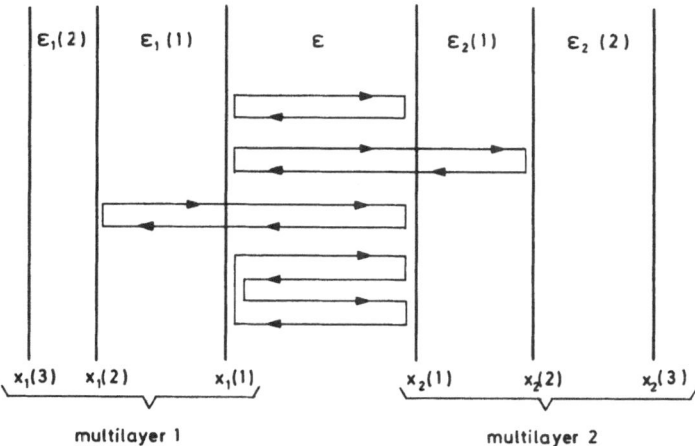

Fig. 23. Contributions to ΔE_{12}

where the order of terms in Eq. (4.92) corresponds to the order of graphs in Fig. 23. The factor l^{-3} in the reflection terms of order l consists of a factor l^{-1}, avoiding multiple counting, and of a factor l^{-2}, accounting for the increasing distance between image charges.

It is obvious from the distance terms in (4.92) that the relative contribution of the inner layers, where $n \geq 2$, to the dispersion energy decreases with decreasing separation $d = x_2(1) - x_1(1)$.

4.8. Continuous Permeability

At zero separation d we obtain an infinite dispersion energy, regardless of whether we consider planar multilayers or layers adsorbed on spheres or cylinders. The use of a constant macroscopic permeability up to the surface is clearly inconsistent with the atomic structure of matter. What is the actual position of the surface relative to the outermost atoms, i.e., what is the minimum distance between the surfaces of the two macroscopic bodies? Reasonable agreement with adhesion measurements is usually obtained by assuming the minimum distance to be approximately 0.5 nm [18].

In the following, we investigate the dispersion energy between spheres whose permeability is a continuous function of the radial coordinate. We shall find that the dispersion energy between two such spheres approaches a finite value at zero distance, if their susceptibilities decrease continuously towards the surface.

If the dielectric permeability ε is a continuous function of position, the electrostatic potential equation becomes

$$\operatorname{div}\varepsilon(r)\operatorname{grad}V(r) = \nabla \cdot \varepsilon(r)\,\nabla V(r) = 0. \tag{4.93}$$

Distinguishing between the variation of $\varepsilon(r)$ and that of $V(r)$ we obtain

$$\nabla \cdot \nabla V(r) + \nabla \ln \varepsilon(r) \cdot \nabla V(r) = 0. \tag{4.94}$$

We find that any variation in $\varepsilon(r)$ causes a local polarization charge which acts as a perturbance for the potential $V(r)$. Let us first consider a single particle j with varying dielectric permeability $\varepsilon_j(r)$ which is embedded in a medium with constant dielectric permeability ε. The reaction potential $V_{ret}(r_k)$ of particle j caused by an external potential $V_{ext}(r_i)$ is conveniently calculated by iteratively applying the integral form of (4.94)

$$V_{ret}(r_k) = (4\pi)^{-1} \int_j dr_i\, \nabla \ln \varepsilon(r_i) \cdot \nabla V(r_i)/|r_i - r_k|. \tag{4.95}$$

The external potential $V_{ext}(r_i)$ induces the field $-\nabla V_{ext}(r_i)$ and the polarization charge $\nabla \ln \varepsilon_j \cdot \nabla V_{ext}(r_i)$, which gives rise to the first order reaction potential $V_{ret}(r_k)$ at position r_k. The first order reaction potential $V_{ret}(r_k)$, in turn, induces a polarization charge and a second order reaction potential, and so on.

Applying Eq. (4.95) to a single sphere 1 with center r_1, we assume the external potential to be the concentric multipole potential

$$u_m^\mu(r_i - r_1) = |r_i - r_1|^m P_m^\mu(\cos \vartheta_i)\exp(i\mu\varphi_i)/(m+\mu)! \tag{4.96}$$

Substituting Eq. (4.96) into Eq. (4.95) and expanding $|r_i - r_k|$ at center r_1 according to Eqs. (4.9), (4.12) and Fig. 12, we obtain

$$V_{ret}(r_k) = -\Delta_1(m, 1)\,|r_k - r_1|^{-(m+1)} P_m^\mu(\cos \vartheta_k)\exp(i\mu\varphi_k)/(m+\mu)! \tag{4.97}$$

where

$$\Delta_1(m, 1) = -m(2m+1)^{-1} \int_0^{R_1} dr\, r^{2m+1}\, d\ln\varepsilon_1(r)/dr. \tag{4.98}$$

The external reaction potential of sphere 1 is the decreasing multipole potential $v_m^\mu(r_k - r_1)$ according to (4.39), with the respective multipole susceptibility given by (4.98). The subscript 1 indicates that we have restricted ourselves to calculating the reaction potential of first order.

For the second order contribution $\Delta_2(m, 1)$ we find

$$\Delta_2(m, 1) = m(2m + 1)^{-2} \int\limits_0^{R_1} dr\, r^{2m+1}(d \ln\varepsilon_1(r)/dr)$$

$$\cdot \int\limits_r^{R_2} ds(d \ln\varepsilon_1(s)/ds).$$

(4.99)

If the dielectric permeability $\varepsilon_1(r)$ of sphere 1 is larger than that in the exterior, we find that $d \ln\varepsilon_1(s)/ds$ is negative across the surface and $\Delta_1(m, 1)$ as well as $\Delta_2(m, 1)$ are positive. The first order and the second order reaction fields exhibit the same sign. The first order reaction field further increases the influence of the exterior field by pushing it out of sphere 1. This applies in a similar manner to all higher order reaction fields. The screening of the interior fields increases the multipole susceptibility of the particle considered.

If, in particular, we assume $\varepsilon_1(r)$ to be independent of r within sphere 1, we reobtain the reaction potential found in Section 4.3 by summing $\Delta_n(m, 1)$ over all orders n.

In order to find the dispersion energy between two spheres 1 and 2 with radially varying dielectric permeabilities $\varepsilon_1(r)$ and $\varepsilon_2(r)$, we use Eq. (4.32), yielding

$$\Delta E_{12} = - (\hbar/4\pi) \int\limits_{-\infty}^{+\infty} d\omega \int\limits_0^{R_1} dr_i(d \ln\varepsilon_1/dr_i) \int\limits_0^{R_2} dr_j(d \ln\varepsilon_2/dr_j)$$

$$\cdot \sum\limits_{m,n=1}^{\infty} \binom{2m+2n}{2m} \frac{m}{2m+1}(r_i/r_{12})^{2m+1} \frac{n}{2n+1}(r_j/r_{12})^{2n+1}.$$

(4.100)

By summing this hypergeometric series over m, n we obtain

$$\Delta E_{12} = - (\hbar/32\pi) \int\limits_{-\infty}^{+\infty} d\omega \int\limits_0^{R_1} dr_i\, d \ln\varepsilon_1/dr_i \int\limits_0^{R_2} dr_j\, d \ln\varepsilon_2/dr_j$$

$$\cdot \left\{ \frac{r_i r_j}{r_{12}^2 - (r_i + r_j)^2} + \frac{r_i r_j}{r_{12}^2 - (r_i - r_j)^2} + \tfrac{1}{2} \ln \frac{r_{12}^2 - (r_i + r_j)^2}{r_{12}^2 - (r_i - r_j)^2} \right\}.$$

(4.101)

If $\varepsilon_j(r)$ changes abruptly across the surface of sphere j, we find that $d \ln\varepsilon_j/dr$ becomes a δ-function at $r = R_j$ and ΔE_{12} varies inversely with the separation $d = r_{12} - R_1 - R_2$ of spheres 1 and 2.

On the other hand, if $\varepsilon_j(r)$ continuously approaches the external permeability ε at the surface of sphere j, we find that $d \ln\varepsilon_j/dr$ assumes a constant value at $r = R_j$ and ΔE_{12} is finite for zero separation $d = r_{12} - R_1 - R_2$. The integral $\int dr_i \int dr_j (r_{12} - r_i - r_j)^{-1} = (r_{12} - r_i - r_j)(\ln(r_{12} - r_i - r_j) - 1)$ remains finite when the upper integration limit $r_{12} - R_1 - R_2$ becomes zero. The divergence of ΔE_{12} with vanishing separation d is clearly a result of taking $\varepsilon_j(r)$ to be a step function across the surface.

5. Retardation

5.1. Radiation

In the preceding investigations we repeatedly touched on the question of retardation. Each fluctuation field of molecule i requires the propagation time r_{ij}/c before it reaches and polarizes molecule j. The reaction field of molecule j is delayed at molecule i for a time $2r_{ij}/c$. This retarded interaction causes a lesser correlation and a smaller energy gain than in the case of an immediate reaction. Similarly, it is inconsistent to consider localized electrostatic fields which oscillate with frequency ω. Here again we have to require that the separation of the particles under consideration is small enough to justify neglecting retardation.

In order to take account of retardation, we have to apply electrodynamics rather than electrostatics. All fields must satisfy the Helmholtz wave equation rather than the Laplace equation.

However, the electrodynamic fields satisfying the Helmholtz equation have a non-vanishing Poynting vector. They loose radiation energy to the exterior. A Hertzian dipole and all Hertzian multipoles are generally damped by outgoing radiation. As a by-product, we find all multipole fields not to be normalizable in infinite space. When applying the above procedure for calculating the dispersion energy, we have to find a means of balancing the outgoing radiation, i.e., to obtain the energy back from infinity.

This balance is achieved by introducing a finite cavity which reflects all waves perfectly. As well as excluding any energy loss, the cavity guarantees normalizability of all waves. Furthermore, it splits up the continuous frequency spectrum of the Hertzian multipoles. It is a discrete set of coupled ingoing and outgoing modes, which interacts with the particles under consideration.

Whereas the coupling of ingoing and outgoing modes and the resulting discrete set of eigenfrequencies depend on the properties and the size of the cavity, this does not apply to the density of states and to the dispersion energy. It is possible to eliminate the properties of the cavity by increasing its size to infinity in the final energy expression. The cavity eventually turns out to be an auxiliary condition for preventing radiation, non-normalizability and the continuous frequency spectrum in the course of integration.

The first investigations into retardation were reported by Casimir and Polder in 1948 [26]. They introduced the randomly fluctuating vector potential of the electromagnetic radiation into the Schrödinger equation and calculated the resulting energy of interaction between two atoms in fourth order perturbation. Any energy loss to the exterior is excluded by assuming periodic boundary conditions. The electric interaction of the atoms under investigation is treated electrostatically, which is permitted within the scope of fourth order perturbation theory: The final energy expression contains the second order of the electric and the fourth order of the magnetic potential.

The retarded dispersion energy between two atoms i and j is found to obey a r_{ij}^{-7} relationship with regard to the separation r_{ij} rather than the nonretarded r_{ij}^{-6} relationship.

Since the investigations based on the integration of pair interactions in Sections 2.4–2.6 cover arbitrary power laws r_{ij}^{-m}, where $m \geq 6$, they apply to nonretarded and retarded interactions as well.

The retarded dispersion energy between macroscopic particles was treated by Liftshitz [28]. He considered half-spaces. Going half the way from the microscopic to the macroscopic approach, Lifshitz expanded the local fluctuations within the half-spaces in terms of plane waves and coupled them to the outgoing (reflected) radiation field. Then, satisfying the boundary conditions for the radiation field across the surfaces of the half-spaces under consideration, he found their force of attraction from Maxwell's stress tensor in the interspace.

This procedure involves some physical difficulties which were intuitively overcome by Lifshitz. For instance he coupled the fluctuations to the outgoing radiation only, which implies damping and causes the total force onto the surface to diverge. Lifshitz cancelled the divergent term in the total force using the argument that it is independent of the separation and is compensated by similar forces on the rear surface of the half-space. In other words, it is compensated by the total force caused by the neglected ingoing radiation. A further consequence of restricting the investigations to the outgoing radiation is the appearance of branch points connecting four different Riemann surfaces of the dispersion function $G(\omega, k)$. These four Riemann surfaces correspond to the different possible combinations of ingoing and outgoing modes at the left-hand and right-hand side. Lifshitz proved that all frequency integrations over the dispersion function $G(\omega, k)$ can be shifted to the imaginary axis without leaving the Riemann surface corresponding to outgoing modes. He ended up with twice the integration along the upper half axis, which means that damping is taken into account rather inadequately.

The proposed cavity, by coupling the amplitudes of the ingoing and outgoing modes rather than their intensities, easily abolishes the difficulties mentioned above. The exact dispersion function $G(\omega, k)$ does not contain any branch point.

The calculation of the force of attraction from Maxwell's stress tensor, i.e. the omission of the Poynting vector in the total energy balance, is equivalent to calculating the dispersion energy merely from the real part of the free energy gain according to Eq. (3.72).

5.2. Transverse Modes

To find the retarded dispersion energy between macroscopic particles, we have to substitute the normalized solutions of the Helmholtz equation for the electrostatic potentials satisfying the Laplace equation. There are generally three types of solutions of the Helmholtz equation: longitudinal modes, which require the existence of free charges and currents and are thus omitted in the following investigations, and transverse modes, which can be subdivided into electric and magnetic modes according to the orientation of the respective fields at particular surfaces.

Let us consider a homogeneous medium with dielectric permeability ε, magnetic permeability μ, and electrical conductivity σ. In the absence of free charges and currents, the Maxwell equations read

$$\partial \boldsymbol{B}/\partial(ct) + \boldsymbol{V} \times \boldsymbol{E} = 0 \qquad \boldsymbol{V} \cdot \boldsymbol{B} = 0 \tag{5.1}$$

$$-\partial \boldsymbol{D}/\partial(ct) + \boldsymbol{V} \times \boldsymbol{H} = \sigma \boldsymbol{E}/c \qquad \boldsymbol{V} \cdot \boldsymbol{D} = 0. \tag{5.2}$$

We satisfy Eq. (5.1) as usual by introducing the four potential (A, V) according to

$$\boldsymbol{B} = \boldsymbol{V} \times \boldsymbol{A} \qquad \boldsymbol{E} = -\partial \boldsymbol{A}/\partial(ct) - \boldsymbol{V} V. \tag{5.3}$$

Equations (5.2) then read

$$\{\boldsymbol{V} \cdot \boldsymbol{V} - \mu \partial/\partial(ct) \, [\varepsilon \partial/\partial(ct) + \sigma/c]\} \, (A, V) = 0 \tag{5.4}$$

with the components of the four potential (A, V) being related by the Lorentz gauge

$$\boldsymbol{V} \cdot \boldsymbol{A} + \mu [\varepsilon \partial/\partial(ct) + \sigma/c] \, V = 0. \tag{5.5}$$

Relation (5.5) reduces to an identity if we introduce a superpotential (Hertzian vector) Z according to

$$A = \mu [\varepsilon \partial/\partial(ct) + \sigma/c] \, Z \qquad V = -\boldsymbol{V} \cdot \boldsymbol{Z}. \tag{5.6}$$

The potential Eq. (5.4) are satisfied if

$$\{\boldsymbol{V} \cdot \boldsymbol{V} - \mu \partial/\partial(ct) \, [\varepsilon \partial/\partial(ct) + \sigma/c]\} \, \boldsymbol{Z} = 0. \tag{5.7}$$

Being generally interested in the harmonic potentials

$$\boldsymbol{Z} = \boldsymbol{Z}_\omega \exp(-i\omega t); (A, V) = (A_\omega, V_\omega) \exp(-i\omega t) \tag{5.8}$$

we find $\boldsymbol{Z}_\omega$ to satisfy the Helmholtz equation

$$\{\boldsymbol{V} \cdot \boldsymbol{V} + K^2\} \, \boldsymbol{Z}_\omega = 0 \tag{5.9}$$

where

$$K^2 = (\omega/c)^2 \, \mu(\varepsilon + i\sigma/\omega) \, . \tag{5.10}$$

The subscript ω is omitted subsequently for notational convenience.

Since the curl operator commutes with the differential operator in Eq. (5.9) we find that

$$\boldsymbol{V} \times \boldsymbol{Z}, \quad \boldsymbol{V} \times \boldsymbol{V} \times \boldsymbol{Z} \tag{5.11}$$

together with $\boldsymbol{Z}$ also satisfy the Helmholtz equation. We obtain up to two additional solutions, which, according to Eqs. (5.3) and (5.6), equal the magnetic and the electric fields. Any further application of the curl operator alternatively reproduces these fields, i.e., we may in particular choose the superpotential to be proportional to the electric field, yielding

$$E = \boldsymbol{V} \times \boldsymbol{V} \times \boldsymbol{Z} = K^2 \boldsymbol{Z} \tag{5.12}$$

$$H = -(i\omega/c)(\varepsilon + i\sigma/\omega)(\boldsymbol{V} \times \boldsymbol{Z}) \tag{5.13}$$

and

$$\boldsymbol{V} \cdot \boldsymbol{Z} = 0; \quad V = 0 \, . \tag{5.14}$$

Complete sets of eigenvectors $\boldsymbol{Z}$ of Helmholtz equation (5.9) are available in spherical, cylindrical and rectangular coordinates [4].

The spherical solutions of the Helmholtz equation can be written

$$Z_s(r) = (K^{-1}\boldsymbol{V} \times)^s r [a_1 j_m(Kr) + a_2 y_m(Kr)]$$

$$\cdot P_m^\mu(\cos \vartheta) \exp(i\mu\varphi)/(m+\mu)! \tag{5.15}$$

with $s = 1$ referring to magnetic modes and $s = 2$ referring to electric modes, and $j_m(Kr)$, $y_m(Kr)$ being spherical Bessel functions of the first or second kind.

Applying cylindrical coordinates we obtain in an analogous manner

$$Z_s(r) = (K^{-1}\boldsymbol{V} \times)^s n_z[a_1 J_m(k\varrho) + a_2 Y_m(k\varrho)] \exp(im\varphi) \exp(ik_z z) \tag{5.16}$$

where $J_m(k\varrho)$, $Y_m(k\varrho)$ are cylindrical Bessel functions of the first or second kind, n_z is the unit vector parallel to the cylinder axis, and k and k_z satisfy the relation

$$k^2 + k_z^2 = K^2 \, . \tag{5.17}$$

In rectangular coordinates we find in like manner

$$Z_s(r) = (K^{-1}\boldsymbol{V} \times)^s n_x [a_1 \exp(ikx)$$

$$+ a_2 \exp(-ikx)] \exp[i(k_y y + k_z z)] \tag{5.18}$$

where n_x is the unit vector in the x direction and k, k_y, k_z satisfy the relationship

$$k^2 + k_y^2 + k_z^2 = K^2 . \tag{5.19}$$

The potentials (5.15), (5.16), (5.18) are obviously built up in like manner as are the respective electrostatic potentials $V(r)$ in Section 4.3–4.5. The scalar potentials used become the electrostatic potentials $u(r)$, $v(r)$ if K approaches zero, i.e. when the velocity of light c approaches infinity.

5.3. Multipole Susceptibilities

Proceeding in a similar manner as in the nonretarded case, we now solve the Helmholtz equation in the presence of a single sphere, cylinder, or half-space j. Let the particle under investigation have the dielectric permeability ε_j, the magnetic permeability μ_j, and the electrical conductivity σ_j. Normalizability of the superpotential within a sphere or cylinder means that no Bessel functions of the second kind are permitted in the interior. In order to obtain normalizability within a half-space, we have to introduce a rear surface, which is situated at the intersection $x = x_s$ of the half-space with the perfectly reflecting cavity. For the coefficients of the internal potentials we find in the case of

spheres: $a_2(\text{int}) = 0$ $\hfill$ (5.20)

cylinders: $a_2(\text{int}) = 0$ $\hfill$ (5.21)

half-spaces: $a_1(\text{int}) \exp(ik_j x_s) + a_2(\text{int}) \exp(-ik_j x_s) = 0 .$ $\hfill$ (5.22)

At the surface of the particle under investigation we require continuity of the normal components of the electric displacement D and of the magnetic induction B and of the tangential components of the electric field E and the magnetic field H. These six boundary conditions reduce to a maximum of four owing to the fact that i) the normal component of the magnetic induction and the tangential component of the electric field, and ii) the normal component of the electric displacement and the tangential components of the magnetic field are linearly dependent on one another according to the Maxwell equations (5.1), (5.2). By linear combinations of the magnetic and electric modes (5.15), (5.16), or (5.18), the remaining four boundary conditions for the tangential components of the electric and magnetic fields can usually be satified. We denote the respective coefficients a_1, a_2 by a second subscript s, so that

$- a_{11}, a_{21}$ refer to magnetic modes, and

$- a_{12}, a_{22}$ refer to electric modes

and obtain four relations between two internal and four external potential coefficients. By eliminating the interior coefficients and solving with respect to the coefficients a_{21}, a_{22} of outgoing modes we find

$$\begin{pmatrix} \Delta_{11}(m,j) & \Delta_{21}(m,j) \\ \Delta_{12}(m,j) & \Delta_{22}(m,j) \end{pmatrix} \begin{pmatrix} a_{11} \\ a_{12} \end{pmatrix} + \begin{pmatrix} a_{21} \\ a_{22} \end{pmatrix} = 0 . \tag{5.23}$$

Equation (5.23) relates the intensities of the outgoing magnetic and electric modes to those of the ingoing modes in a similar manner to the scalar equations (4.42), (4.52), and (4.69) in the nonretarded limit.

For the multipole susceptibilities $\Delta_{11}(m,j)$, $\Delta_{21}(m,j)$, $\Delta_{12}(m,j)$, and $\Delta_{22}(m,j)$ we find in the case of a sphere j

$$\Delta_{11}(m,j) = \frac{\mu_j j_m(K_j R_j)[K R_j j_m(K R_j)]' - \mu j_m(K R_j)[K_j R_j j_m(K_j R_j)]'}{\mu_j j_m(K_j R_j)[K R_j y_m(K R_j)]' - \mu y_m(K R_j)[K_j R_j j_m(K_j R_j)]'} \tag{5.24}$$

$$\Delta_{22}(m,j) = [\text{interchange } \varepsilon_j \text{ and } \mu_j, \varepsilon \text{ and } \mu \text{ in Eq. (5.24)}] \tag{5.25}$$

$$\Delta_{12}(m,j) = \Delta_{21}(m,j) = 0 . \tag{5.26}$$

In the case of spheres there is no coupling between electric and magnetic modes as a result of applying the boundary conditions. The magnetic modes react to the change $\mu_j \leftrightarrow \mu$ of the magnetic permeability, the electric modes react to the change $\varepsilon_j \leftrightarrow \varepsilon$ of the electric permeability. Here and in the following we include the damping caused by the currents in ε_j, i.e., we understand ε_j to equal the effective permeability $\varepsilon_j + i\sigma_j/\omega$.

In the presence of a cylinder j we obtain

$$\varkappa\Delta_{11}(m,j) = \begin{vmatrix} k_j R_j J_m(k_j R_j) & k R_j J_m(k R_j) \\ \mu_j [J_m(k_j R_j)]' & \mu[J_m(k R_j)]' \end{vmatrix} \begin{vmatrix} k_j R_j J_m(k_j R_j) & k R_j Y_m(k R_j) \\ \varepsilon_j [J_m(k_j R_j)]' & \varepsilon[Y_m(k R_j)]' \end{vmatrix}$$
$$- [m k_z(\omega/c)^{-1}(k/k_j - k_j/k) J_m(k_j R_j)]^2 J_m(k R_j) Y_m(k R_j) \tag{5.27}$$

$$\varkappa\Delta_{22}(m,j) = [\text{interchange } \varepsilon_j \text{ and } \mu_j, \varepsilon \text{ and } \mu \text{ in Eq. (5.27)}] \tag{5.28}$$

$$\varkappa\Delta_{12}(m,j) = \varkappa\Delta_{21}(m,j) = 2\pi^{-1} K m k_z k^{-2}(\varepsilon\mu - \varepsilon_j\mu_j) J_m^2(k_j R_j) \tag{5.29}$$

where

$$\varkappa = \begin{vmatrix} k_j R_j J_m(k_j R_j) & k R_j Y_m(k R_j) \\ \mu_j [J_m(k_j R_j)]' & \mu[Y_m(k R_j)]' \end{vmatrix} \begin{vmatrix} k_j R_j J_m(k_j R_j) & k R_j Y_m(k R_j) \\ \varepsilon_j [J_m(k_j R_j)]' & \varepsilon[Y_m(k R_j)]' \end{vmatrix}$$
$$- [m k_z(\omega/c)^{-1}(k/k_j - k_j/k) J_m(k_j R_j)]^2 Y_m^2(k R_j) . \tag{5.30}$$

There is a symmetric coupling between electric and magnetic modes. In the nonretarded limit, when c approaches infinity, we find that the second terms in Eqs. (5.27), (5.28), and (5.30) vanish and the magnetic and electric modes respond to the changes $\mu_j \leftrightarrow \mu$ and $\varepsilon_j \leftrightarrow \varepsilon$, respectively.

For the electric multipole susceptibility $\varDelta_{22}(m,j)$ we again obtain the non-retarded expression (4.53).

Evaluation of boundary conditions in the presence of a half-space j yields

$$\varDelta_{11}(k,j) = \exp(2ikx_j)\frac{k_j\mu_j^{-1}\coth k_j(x_j-x_S) - ik\mu^{-1}}{k_j\mu_j^{-1}\coth k_j(x_j-x_S) + ik\mu^{-1}} \tag{5.31}$$

$$\varDelta_{22}(k,j) = \exp(2ikx_j)\frac{k_j\varepsilon_j^{-1}\coth k_j(x_j-x_S) - ik\varepsilon^{-1}}{k_j\varepsilon_j^{-1}\coth k_j(x_j-x_S) + ik\varepsilon^{-1}} \tag{5.32}$$

$$\varDelta_{12}(k,j) = \varDelta_{21}(k,j) = 0. \tag{5.33}$$

There is no coupling of electric and magnetic modes via the boundary conditions.

5.4. Transposition

Turning to the interaction between two particles 1 and 2, the main problem is to find the appropriate addition theorem for the potentials used. The outgoing modes of particle 1 are the ingoing modes for particle 2, and vise versa. We must expand the outgoing modes of particle 1 in terms of incoming modes at particle 2. This is generally possible: Any complete set of eigenvectors of a linear differential operator can be expanded in terms of any other complete set. All the fields and potentials under investigation satisfy the Helmholtz equation and have vanishing divergence: Modes localized at different positions r_1 and r_2 can be expanded in terms of each other.

It is again appropriate to use inversely oriented coordinates as shown in Figs. 15 and 17. The addition theorem valid for the spherical potentials (5.15) is found by repeated differentiation of the addition theorem for spherical Bessel functions of zero order, Eqs. (10.1.45) and (10.1.46) in

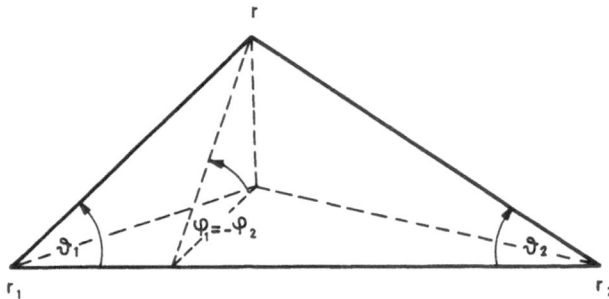

Fig. 24. Inverted spherical coordinates

Ref. [1], with respect to separation and angles. This is described in full detail in Section 5.6. If $f_m(\zeta)$ is a spherical Bessel function $j_m(\zeta)$, $y_m(\zeta)$ of the first or the second kind, we obtain (see Fig. 24)

$$
f_m(K|\mathbf{r} - \mathbf{r}_2|)\, P_m^\mu(\cos\vartheta_2)\exp(i\mu\varphi_2)
$$

$$
= (-1)^\mu \sum_{n=\mu}^{\infty} (2n+1)\, U_{mn}^\mu(Kr_{21})\, j_n(K|\mathbf{r}-\mathbf{r}_1|)\, P_n^{-\mu}(\cos\vartheta_1)\exp(-i\mu\varphi_1)
$$

(5.34)

where

$$
U_{mn}^\mu(\zeta) = (2/\zeta)^\mu \sum_{v=0}^{m-\mu} (-1)^v \frac{\Gamma(m-v+\tfrac{1}{2})\,\Gamma(n-v+\tfrac{1}{2})\,\Gamma(\mu+v+\tfrac{1}{2})}{\Gamma(m+n-\mu-v+\tfrac{3}{2})\,\Gamma(\mu+\tfrac{1}{2})\,\Gamma(\tfrac{1}{2})}
$$

$$
\cdot \frac{(m+n-v)!}{(m-\mu-v)!\,(n-\mu-v)!\,v!}(m+n-\mu-2v+\tfrac{1}{2})\,f_{m+n-\mu-2v}(\zeta).
$$

(5.35)

Applying this scalar addition theorem to the magnetic and electric potentials $Z_s(\mathbf{r} - \mathbf{r}_j)$ with $s = 1, 2$ we find:

$$
(K^{-1}\boldsymbol{\nabla}\times)^s(\mathbf{r}-\mathbf{r}_2)\, f_m(K|\mathbf{r}-\mathbf{r}_2|)\, P_m^\mu(\cos\vartheta_2)\exp(i\mu\varphi_2)
$$

$$
= (-1)^\mu \sum_{n=\mu}^{\infty} (2n+1)\,\{V_{mn}^\mu(Kr_{21}) + W_{mn}^\mu(Kr_{21})(K^{-1}\boldsymbol{\nabla}\times)\}
$$

(5.36)

$$
\cdot (K^{-1}\boldsymbol{\nabla}\times)^s(\mathbf{r}-\mathbf{r}_1)\, j_n(K|\mathbf{r}-\mathbf{r}_1|)\, P_n^{-\mu}(\cos\vartheta_1)\exp(-i\mu\varphi_1)
$$

where

$$
V_{mn}^\mu(\zeta) = U_{mn}^\mu(\zeta) - \frac{n-\mu+1}{(n+1)(2n+1)}\zeta U_{m\,n+1}^\mu(\zeta) - \frac{n+\mu}{n(2n+1)}\zeta U_{m\,n-1}^\mu(\zeta)
$$

(5.37)

$$
W_{mn}^\mu(\zeta) = i\mu[n(n+1)]^{-1}\zeta\, U_{mn}^\mu(\zeta).
$$

(5.38)

Addition theorem (5.36) obviously couples magnetic and electric modes.

In the following we distinguish between the coefficients $V_{mn}^\mu(\zeta)$, $W_{mn}^\mu(\zeta)$ which arise on transposing spherical Bessel functions $j_m(\zeta)$, $y_m(\zeta)$ of the first or the second kind by adding a second argument $V_{mn}^\mu(\zeta,j)$, $W_{mn}^\mu(\zeta,j)$, where $j = 1, 2$.

The addition theorem for the cylindrical potentials $Z_s(\mathbf{r} - \mathbf{r}_j)$ according to Eq. (5.16) is readily obtained from Grafs addition theorem, Eq. (9.1.79) in Ref. [1]. Using the inverted cylindrical coordinates shown in Fig. 17 we have

$$
J_m(k\varrho_2)\exp(im\varphi_2) = \sum_{n=-\infty}^{+\infty} J_{m+n}(k r_{21})\, J_n(k\varrho_1)\exp(in\varphi_1)
$$

(5.39)

$$
Y_m(k\varrho_2)\exp(im\varphi_2) = \sum_{n=-\infty}^{+\infty} Y_{m+n}(k r_{21})\, J_n(k\varrho_1)\exp(in\varphi_1).
$$

(5.40)

Addition theorems (5.39), (5.40) also apply in an identical manner to the potentials $Z_s(r - r_j)$. There is no mixing of electric and magnetic modes by the transposition of potentials in the case of cylinders.

The transposition of the planar potentials $Z_s(r - r_j)$ according to Eq. (5.18) obviously renders the phase factors $\exp[\pm ik(x_2 - x_1)]$.

5.5. Pair States

We are now well prepared to calculate the localized modes in the presence of two particles 1 and 2. We construct them from normalized electric and magnetic modes $s = 1, 2$ localized at the individual particles $j = 1, 2$. In the case of spheres we put

$$
\begin{aligned}
Z(r) = \sum_{j=1,2} \sum_{s=1,2} (K^{-1} V \times)^s (r - r_j) \sum_{m=\mu}^{\infty} \{a_{1s}(m,j) j_m(K|r - r_j|) \\
+ a_{2s}(m,j) y_m(K|r - r_j|)\} P_m^{\pm\mu}(\cos\vartheta_j) \exp(\pm i\mu\varphi_j)/(m \pm \mu)!
\end{aligned}
\tag{5.41}
$$

The mutual inversion of coordinates at spheres 1 and 2 entails accounting for the rotational symmetry by coupling inverted spherical harmonics $P_m^\mu(\cos\vartheta_1) \exp(i\mu\varphi_1)$ and $P_m^{-\mu}(\cos\vartheta_2) \exp(-i\mu\varphi_2)$. Normalization of all modes used is obtained if we require the normal components of all fields to vanish on the surface of a perfectly reflecting cavity with radius r_S, yielding

$$
a_{1s}(m,j) j_m(Kr_S) + a_{2s}(m,j) y_m(Kr_S) = 0 .
\tag{5.42}
$$

By transposing all modes to a single sphere i by means of addition theorem (5.36) and satisfying boundary condition (5.23) we obtain

$$
\begin{aligned}
\Delta_{ss}(m,i) \left[a_{1s}(m,i) + (2m+1)(-1)^\mu \sum_{n=\mu}^{\infty} [(m \pm \mu)!/(n \mp \mu)!] \right. \\
\left. \cdot \{a_s(n,j) V_{nm}^{\mp\mu}(Kr_{21}) + a_t(n,j) W_{nm}^{\mp\mu}(Kr_{21})\} \right] + a_{2s}(m,i) = 0
\end{aligned}
\tag{5.43}
$$

where

$$
j = \begin{cases} 2 \\ 1 \end{cases} \text{ for } i = \begin{cases} 1 \\ 2 \end{cases}; \quad t = \begin{cases} 2 \\ 1 \end{cases} \text{ for } s = \begin{cases} 1 \\ 2 \end{cases}
\tag{5.44}
$$

and

$$
\begin{aligned}
a_s V_{nm}^\mu(\zeta) = a_{1s} V_{nm}^\mu(\zeta, 1) + a_{2s} V_{nm}^\mu(\zeta, 2) \\
a_s W_{nm}^\mu(\zeta) = a_{1s} W_{nm}^\mu(\zeta, 1) + a_{2s} W_{nm}^\mu(\zeta, 2)
\end{aligned}
\tag{5.45}
$$

Conditions (5.42), (5.43) fully determine the amplitudes and frequencies of the localized electromagnetic modes in the presence of two spheres 1 and 2. We have built up $Z(r)$ at each sphere 1 and 2 from ingoing and outgoing electric and magnetic modes, i.e., each multipole brings about eight different amplitudes $a_{11}(m, 1), \ldots, a_{22}(m, 2)$. The cavity imposes the four conditions (5.42) and each sphere imposes the two conditions (5.43) on these eight amplitudes. The eigenfrequencies of the localized modes under investigation are the eigenvalues of the secular determinant resulting from Eqs. (5.42), (5.43).

Similarly, to find the pair states in the presence of two cylinders 1 and 2 we put

$$
\mathbf{Z}(\mathbf{r}) = \sum_{j=1,2} \sum_{s=1,2} (K^{-1} \mathbf{V} \times)^s \mathbf{n}_z \exp(i k_z z)
$$

$$
\cdot \sum_{m=-\infty}^{+\infty} \{a_{1s}(m, j) J_m(k \varrho_j) + a_{2s}(m, j) Y_m(k \varrho_j)\} \exp(i m \varphi_j) . \tag{5.46}
$$

A perfectly reflecting cavity with radius r_S imposes the normalization condition

$$
a_{1s}(m, j) J_m(k r_S) + a_{2s}(m, j) Y_m(k r_S) = 0 . \tag{5.47}
$$

On transposition of all modes to cylinder i by means of addition theorems (5.39), (5.40) and by application of boundary condition (5.23), we obtain

$$
\sum_{t=1,2} \Delta_{ts}(m, i) \left[a_{1t}(m, i) + \sum_{n=-\infty}^{+\infty} \{a_{1t}(n, j) J_{m+n}(k r_{21}) \right.
$$

$$
\left. + a_{2t}(n, j) Y_{m+n}(k r_{21})\} \right] + a_{2s}(m, i) = 0 \tag{5.48}
$$

with

$$
j = \begin{cases} 2 \\ 1 \end{cases} \quad \text{for} \quad i = \begin{cases} 1 \\ 2 \end{cases} . \tag{5.49}
$$

The symmetry of the two cylinders with respect to a reflection at any plane $\mathbf{n}_z$ permits a further splitting of the secular system (5.48). Using the fact that the cylindrical Bessel functions $J_m(\zeta)$, $Y_m(\zeta)$ are odd functions with respect to order, we find

$$
a_{ts}(-m, i) = \pm (-1)^{m+s} a_{ts}(m, i) . \tag{5.50}
$$

The electric and the magnetic modes show the inverse symmetry behavior on reflection at the planes $\mathbf{n}_z$. Eqs. (5.50) can be used for reducing Eq. (5.48) to positive values of m, n.

Similarly to the case of spheres, we find a coupling of ingoing and outgoing electric and magnetic modes at each cylinder 1 and 2. The eigenfrequencies of the resulting localized modes are the eigenvalues of the secular determinant corresponding to Eqs. (5.47) and (5.48).

The simplest case is that of two half-spaces 1 and 2. Neither the boundary condition (5.23), nor the addition theorem, couple electric and magnetic modes. The ingoing modes at half-space 1 are identical with the outgoing modes at half-space 2, and vice versa. Normalization of all modes is already achieved by introducing a perfectly reflecting rear surface at $x = x_S$. Putting

$$\mathbf{Z}_s(\mathbf{r}) = (K^{-1} \mathbf{V} \times)^s \mathbf{n}_x \{a_{1s} \exp(ikx) + a_{2s} \exp(-ikx)\} \exp[i(k_y y + k_z z)]$$

$$(5.51)$$

we obtain the boundary conditions

$$\left.\begin{array}{l} \varDelta_{ss}(k,\,1)\,a_{1s} + a_{2s} = 0 \\ a_{1s} + \varDelta_{ss}(k,\,2)\,a_{2s} = 0 \end{array}\right\}. \qquad (5.52)$$

The eigenfrequencies of the localized electromagnetic modes in the presence of two half-spaces are found from the corresponding secular determinant

$$\varDelta_{ss}(k,\,1)\,\varDelta_{ss}(k,\,2) - 1 = 0. \qquad (5.53)$$

5.6. Spherical Addition Theorems

It is the aim of this section to prove the addition theorems (5.34) and (5.36) for the spherical eigenvectors (5.15) of the Helmholtz equation. They are fundamental for treating the retarded interaction potential of spheres and spherical atoms as well.

Let us consider the triangle shown in Fig. 25, and let us take b and β to be functions of the free variables a, c, α. Then, from

$$a \sin \alpha = b \sin \beta; \qquad a \cos \alpha + b \cos \beta = c \qquad (5.54)$$

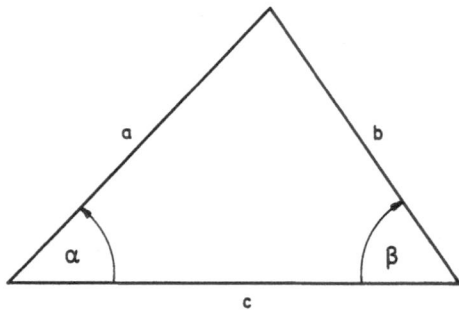

Fig. 25. Sine and cosine theorems

we have

$$\partial b/\partial c = \cos\beta; \quad b\partial\beta/\partial c = -\sin\beta \tag{5.55}$$

and

$$\partial b/\partial\alpha = c\sin\beta; \quad b\partial\beta/\partial\alpha = c\cos\beta - b. \tag{5.56}$$

We are interested in the addition theorem for the functions $f_m(b)\,P_m^\mu(\cos\beta)$ where $f_m(b)$ is a spherical Bessel function $j_m(b)$ or $y_m(b)$ of the first or the second kind. Differentiating with respect to c and making use of Eqs. (10.1.19) and (10.1.20) in Ref. [1] we find

$$(\partial/\partial c)\,[f_m(b)\,P_m^\mu(\cos\beta)] = \cos\beta(\mathrm{d}f_m(b)/\mathrm{d}b)\,P_m^\mu(\cos\beta)$$
$$+ f_m(b)\,b^{-1}\sin^2\beta\,\mathrm{d}P_m^\mu(\cos\beta)/\mathrm{d}\cos\beta \tag{5.57}$$

and

$$(\partial/\partial c)\,[f_m(b)\,P_m^\mu(\cos\beta)] = (2m+1)^{-1}\,\{f_{m-1}(b)\,[\sin^2\beta\,\partial/\partial\cos\beta + m\cos\beta]$$
$$+ f_{m+1}(b)\,[\sin^2\beta\,\partial/\partial\cos\beta - (m+1)\cos\beta]\}\,P_m^\mu(\cos\beta) \tag{5.58}$$

Application of Eqs. (8.5.3) and (8.5.4) in Ref. [1] for the associated Legendre functions $P_m^\mu(\cos\beta)$ on the cut yields

$$(\partial/\partial c)\,[f_m(b)\,P_m^\mu(\cos\beta)] = \frac{m+\mu}{2m+1}f_{m-1}(b)\,P_{m-1}^\mu(\cos\beta)$$
$$- \frac{m-\mu+1}{2m+1}f_{m+1}(b)\,P_{m+1}^\mu(\cos\beta). \tag{5.59}$$

The differentiation with respect to c renders recurrence relation (5.59) with respect to m.

Differentiating with respect to α it is sufficient to consider $\mu = m$. From

$$(\partial/\partial\alpha)\,[f_m(b)\,P_m^m(\cos\beta)] = c\sin\beta(\mathrm{d}f_m(b)/\mathrm{d}b)\,P_m^m(\cos\beta)$$
$$+ f_m(b)\,[(c/b)\cos\beta - 1]\,m\cot\beta\,P_m^m(\cos\beta) \tag{5.60}$$

and

$$\cot\beta(c\cos\beta - b) = b\cot\alpha - c\sin\beta \tag{5.61}$$

we find by using Eqs. (10.1.22) and (8.5.5) in Ref. [1]

$$(\partial/\partial\alpha)\,[f_m(b)\,P_m^m(\cos\beta)] = m\cot\alpha\,f_m(b)\,P_m^m(\cos\beta)$$
$$+ c(2m+1)^{-1}f_{m+1}(b)\,P_{m+1}^{m+1}(\cos\beta). \tag{5.62}$$

The differentiation with respect to α renders recurrence relation (5.62) with respect to $\mu = m$.

Recurrence relations (5.59) and (5.62) enable the addition theorem for $f_m(b) P_m^\mu(\cos\beta)$ with m, μ arbitrary to be found from the respective addition theorem for the spherical Bessel functions $j_0(b)$, $y_0(b)$ of zero order. Eqs. (10.1.45) and (10.1.46) in Ref. [1] give

$$f_0(b) = \sum_{n=0}^{\infty} (2n+1) f_n(c) j_n(a) P_n(\cos\alpha) . \tag{5.63}$$

First we apply recurrence relation (5.62) and obtain

$$f_m(b) P_m^m(\cos\beta) = \prod_{\mu=0}^{m-1} (2\mu+1) c^{-1}$$
$$\cdot (\partial/\partial\alpha - \mu\cot\alpha) \sum_{n=0}^{\infty} (2n+1) f_n(c) j_n(a) P_n(\cos\alpha) . \tag{5.64}$$

Differentiation formula (8.6.6) in Ref. [1] for the associated Legendre functions $P_m^\mu(\cos\alpha)$ gives:

$$f_m(b) P_m^m(\cos\beta) = [(2m)!/2m! \, c^m] \sum_{n=m}^{\infty} (2n+1) f_n(c) j_n(a) P_n^m(\cos\alpha) . \tag{5.65}$$

Recurrence relation (5.59) raises m for fixed μ. It does not explicitly contain the variables a and α, thus changing the dependence of the right hand terms in Eq. (5.65) on c only. Introducing the general relationship

$$f_m(b) P_m^\mu(\cos\beta) = \sum_{n=\mu}^{\infty} (2n+1) [(n-\mu)!/(n+\mu)!] U_{mn}^\mu(c) j_n(a) P_n^\mu(\cos\alpha) \tag{5.66}$$

where

$$U_{mn}^m(c) = [(n+m)!/(n-m)!] [(2m)!/2^m m! \, c^m] f_n(c) \tag{5.67}$$

according to Eq. (5.65) we have from Eq. (5.59)

$$dU_{mn}^\mu(c)/dc = [(m+\mu)/(2m+1)] U_{m-1\,n}^\mu(c)$$
$$- [(m-\mu+1)/(2m+1)] U_{m+1\,n}^\mu(c) . \tag{5.68}$$

Recurrence relation (5.68) and the initial value (5.67) are satisfied if

$$U_{mn}^\mu(c) = (2/c)^\mu \sum_{v=0}^{m-\mu} (-1)^v \frac{\Gamma(m-v+\tfrac{1}{2}) \Gamma(n-v+\tfrac{1}{2}) \Gamma(\mu+v+\tfrac{1}{2})}{\Gamma(m+n-\mu-v+\tfrac{3}{2}) \Gamma(\mu+\tfrac{1}{2}) \Gamma(\tfrac{1}{2})}$$
$$\cdot \frac{(m+n-v)!}{(m-\mu-v)!(n-\mu-v)! \, v!} (m+n-\mu-2v+\tfrac{1}{2}) f_{m+n-\mu-2v}(c) . \tag{5.69}$$

Addition theorem (5.66) turns into addition theorem (5.34) if the notation shown in Fig. 24 is used.

Turning to the transposition of the vector potentials $Z_s(r - r_j)$ according to Eq. (5.15) from center r_2 to center r_1, we apply Eq. (5.34) to the transposition of the scalar factor $f_m(Kr) P_m^\mu(\cos\vartheta) \exp(i\mu\varphi)$. The transposition of r is readily achieved according to $r - r_2 = (r - r_1) - (r_2 - r_1)$. However, we are left with the problem of splitting $V \times (r_2 - r_1) f_m(Kr) P_m(\cos\vartheta) \exp(i\mu\varphi)$ into an electric and a magnetic contribution again. Thus, let us try to satisfy

$$V \times [n f(r) P_m^\mu(\cos\vartheta) \exp(i\mu\varphi)]$$
$$= V \times [r g_1(r)] + (V \times)^2 [r g_2(r)] \tag{5.70}$$

where n is the unit vector in the direction of the polar axis

$$n_r = \cos\vartheta, \quad n_\vartheta = -\sin\vartheta, \quad n_\varphi = 0 \tag{5.71}$$

and $g_1(r)$, $g_2(r)$ are arbitrary functions of r. By representing Eq. (5.70) in polar coordinates we find

$$r^{-1}(\partial/\partial\varphi, \cot\vartheta\,\partial/\partial\varphi, -(\partial/\partial r)r\sin\vartheta - (\partial/\partial\vartheta)\cos\vartheta)$$

$$\cdot f(r) P_m^\mu(\cos\vartheta) \exp(i\mu\varphi) \tag{5.72}$$

$$= (0, (\sin\vartheta)^{-1}\,\partial/\partial\varphi, -\partial/\partial\vartheta) g_1(r)$$

$$+ (-(r\sin\vartheta)^{-2}\{(\sin\vartheta\partial/\partial\vartheta)^2 + (\partial/\partial\varphi)^2\}, \partial^2/\partial r\partial\vartheta, (\sin\vartheta)^{-1}\partial^2/\partial r\partial\varphi) r g_2(r).$$

Equating to zero the radial component we obtain

$$g_2(r) = [m(m+1)]^{-1} (\partial/\partial\varphi) f(r) P_m^\mu(\cos\vartheta) \exp(i\mu\varphi) \tag{5.73}$$

whereas equating to zero the ϑ, φ components yields

$$g_1(r) = r^{-1}(\cos\vartheta - [m(m+1)]^{-1} \sin\vartheta(\partial^2/\partial r\partial\vartheta)r)$$

$$\cdot f(r) P_m^\mu(\cos\vartheta) \exp(i\mu\varphi). \tag{5.74}$$

By applying Eq. (5.70) in particular to the spherical Bessel functions $f_m(r) = j_m(r)$, $y_m(r)$ we obtain

$$g_1(r) = \left\{ \frac{m+\mu}{m(2m+1)} f_{m-1}(r) P_{m-1}^\mu(\cos\vartheta) + \frac{m-\mu+1}{(m+1)(2m+1)} \right.$$

$$\left. \cdot f_{m+1}(r) P_{m+1}^\mu(\cos\vartheta) \right\} \exp(i\mu\varphi). \tag{5.75}$$

Using Eqs. (5.34), (5.70), (5.73), (5.75), we obtain addition theorem (5.36) for the vector potentials $Z_s(r - r_j)$.

6. Retarded Dispersion Energy

6.1. Spheres

The retarded dispersion energy between the particles under investigation is readily obtained by applying the integration technique described in Section 3.4. This method is now rigorous. The contradiction of fluctuating electrostatic fields is removed. For the dispersion function $G(\omega)$ we have to use the ratio of secular determinants obtained in Section 5.5 for finite and infinite separation of the particles considered.

The substantial advantages of this integration technique are the shifting the frequency integration to the imaginary axis and the fact that the eigenfrequencies of the dispersion relation do not have to be calculated explicitly. We need the dispersion function $G(\omega)$ for imaginary frequencies only. Since the ingoing and the outgoing radiation turn into exponentially decreasing modes for purely imaginary values of the frequency, we may turn to the limit of infinite cavities without further affecting the final result.

Let us first consider the attraction between spheres. According to Eq. (5.10) we find the argument Kr of the spherical Bessel functions (5.15) to be imaginary for imaginary frequencies. Both $j_m(Kr)$ and $y_m(Kr)$ increase exponentially with increasing r, i.e. the boundary condition (5.42) caused by the cavity with radius r_S turns into

$$\lim_{r_S \to \infty} a_{2s}(m,j)/a_{1s}(m,j) = -\lim_{r_S \to \infty} j_m(Kr_S)/y_m(Kr_S)$$

$$= \pm i \quad \text{for} \quad \text{Im}(k) \gtrless 0. \tag{6.1}$$

Inserting Eq. (6.1) into the potential (5.41) yields

$$a_{1s}(m,j)\,j_m(Kr) + a_{2s}(m,j)\,y_m(Kr) \tag{6.2}$$

$$= a_{1s}(m,j) \begin{cases} h_m^{(1)}(Kr) \\ h_m^{(2)}(Kr) \end{cases} \quad \text{for} \quad \text{Im}(k) \gtrless 0.$$

We are left with the spherical Bessel functions $h_m^{(1)}(Kr)$ and $h_m^{(2)}(Kr)$ of the third kind, which decrease exponentially with increasing r in the upper and lower half-plane, respectively. This suggests the use of spherical Bessel functions of the third kind from the start. However, in that case it is difficult to interpret the physical relevance of the electromagnetic modes under consideration. They are normalized at imaginary rather than real frequencies.

The exponential decrease of the potential considered similarly affects the coefficients $V_{mn}^\mu(Kr_{21})\,W_{mn}^\mu(Kr_{21})$ arising in the addition theorem (5.36). The relevant expressions (5.45) entering boundary

Table 2. Spherical coupling parameters

m	n	μ	v_k					w_k			
dipole	dipole										
1	1	0	0	1	1			0	0		
		1	1	1	1			1	1		
quadrupole	dipole										
2	1	0	0	1	3	3		0	0	0	
		1	1	3	6	6		1	3	3	
octupole	dipole										
3	1	0	0	1	6	15	15	0	0	0	0
		1	1	6	21	45	45	1	6	15	15
quadrupole	quadrupole										
2	2	0	0	1	5	12	12	0	0	0	0
		1	1	5	21	48	48	1	5	12	12
		2	0	8	24	48	48	0	8	24	24

conditions (5.43) decrease exponentially with increasing separation r_{21}. We obtain

$$V_{mn}^{\mu}(i\zeta) = i^{-(m+n+2)} \binom{m+1}{2} \zeta^{-1} e^{-\zeta} \sum_{k=0}^{m+n} v_k \zeta^{-k} \tag{6.3}$$

$$W_{mn}^{\mu}(i\zeta) = i^{-(m-n)} \binom{m+1}{2} \zeta^{-1} e^{-\zeta} \sum_{k=0}^{m+n-1} w_k \zeta^{-k} . \tag{6.4}$$

The coefficients v_k, w_k necessary for treating interactions up to dipole-octupole and quadrupole-quadrupole are given in Table 2.

Summing over all rotation wave numbers μ we obtain

$$\Delta E_{12} = (\hbar/4\pi i) \int_{-i\infty}^{+i\infty} d\omega \coth(\hbar\omega/2kT) \sum_{\mu=-\infty}^{+\infty} \ln G(\omega,\mu) \tag{6.5}$$

where

$$G(\omega,\mu) = \tag{6.6}$$

$$\det \begin{vmatrix} 1 & 0 & \frac{(2m+1)\Delta_{11}(m,1)}{\Delta_{11}(m,1)+i} V_{nm}^{-\mu} & \frac{(2m+1)\Delta_{11}(m,1)}{\Delta_{11}(m,1)+i} W_{nm}^{-\mu} \\ 0 & 1 & \frac{(2m+1)\Delta_{22}(m,1)}{\Delta_{22}(m,1)+i} W_{nm}^{-\mu} & \frac{(2m+1)\Delta_{22}(m,1)}{\Delta_{22}(m,1)+i} V_{nm}^{-\mu} \\ \frac{(2n+1)\Delta_{11}(n,2)}{\Delta_{11}(n,2)+i} V_{mn}^{\mu} & \frac{(2n+1)\Delta_{11}(n,2)}{\Delta_{11}(n,2)+i} W_{mn}^{\mu} & 1 & 0 \\ \frac{(2n+1)\Delta_{22}(n,2)}{\Delta_{22}(n,2)+i} W_{mn}^{\mu} & \frac{(2n+1)\Delta_{22}(n,2)}{\Delta_{22}(n,2)+i} V_{mn}^{\mu} & 0 & 1 \end{vmatrix}$$

The dispersion functions $G(\omega, \mu)$ and $G(\omega, -\mu)$ are identical, the respective eigenvectors differ by inversion of the magnetic modes relative to the electric modes. The electric and the magnetic modes exhibit the inverse symmetry behavior on changing the sense of rotation. The respective coefficients $V_{mn}^{\mu}(\zeta)$, $W_{mn}^{\mu}(\zeta)$ in addition theorem (5.36) satisfy the relation

$$[(n+\mu)!/(m-\mu)!] \, (V_{m\,n}^{-\mu}(\zeta), \, W_{m\,n}^{-\mu}(\zeta))$$
$$= [(n-\mu)!/(m+\mu)!] \, (V_{mn}^{\mu}(\zeta), \, -W_{mn}^{\mu}(\zeta)) \tag{6.7}$$

as can be seen from Eqs. (5.34), (5.37), (5.38) and from the corresponding behavior of the associated Legendre functions $P_m^{\mu}(\cos\vartheta)$, Eq. (8.2.5) in Ref. [1].

All off-diagonal elements in Eq. (6.6) decrease exponentially with increasing imaginary frequency. By expanding $\ln G(\omega, \mu)$ with respect to the off-diagonal elements up to quadratic terms we obtain

$$\Delta E_{12} = - (\hbar/4\pi i) \int_{-i\infty}^{+i\infty} d\omega \coth(\hbar\omega/2kT) \sum_{s=1,2} \sum_{m,n=1}^{\infty} \frac{(2m+1)\Delta_{ss}(m,1)}{\Delta_{ss}(m,1)+i}$$
$$\cdot \left\{ \frac{(2n+1)\Delta_{ss}(n,2)}{\Delta_{ss}(n,2)+i} \sum_{\mu} V_{mn}^{\mu} V_{nm}^{-\mu} + \frac{(2n+1)\Delta_{tt}(n,2)}{\Delta_{tt}(n,2)+i} \sum_{\mu} W_{mn}^{\mu} W_{nm}^{-\mu} \right\} \tag{6.8}$$

with $t = 2$ for $s = 1$, and vice versa. We find an electric, a magnetic, and a mixed contribution to the total dispersion energy. For the dipole-dipole interaction terms $m = n = 1$ we obtain from Eqs. (6.3), (6.4)

$$\sum_{\mu} V_{11}^{\mu} \, V_{11}^{-\mu} = \tfrac{1}{2}\zeta^{-2} e^{-2\zeta} [1 + 2\zeta^{-1} + 5\zeta^{-2} + 6\zeta^{-3} + 3\zeta^{-4}] \tag{6.9}$$

$$\sum_{\mu} W_{11}^{\mu} \, W_{11}^{-\mu} = - \tfrac{1}{2}\zeta^{-2} e^{-2\zeta} [1 + 2\zeta^{-1} + \zeta^{-2}]. \tag{6.10}$$

$i\zeta = Kr_{21}$, which proves that the electric dipole-dipole term $m = n = 1$, $s = 2$ agrees with the findings reported by Casimir and Polder on the retarded dispersion energy between atoms [26]. The contributions resulting from the interaction between electric and magnetic modes tend to decrease the attraction caused by the interaction of electric modes alone or of magnetic modes alone.

The convergence of all multipole interaction terms in Eq. (6.8) is guaranteed by the decrease of the multipole susceptibilities $\Delta_{ss}(m, j)$, on the one hand, and by the decrease of the phase shift factor $\exp(-2\zeta)$, $i\zeta = Kr_{21}$, with increasing imaginary frequency, on the other hand. At small separations $d = r_{21} - R_1 - R_2$ of the spheres considered we find that the multipole susceptibilities decrease more rapidly than the phase shift factor. Neglecting the latter, we obtain the nonretarded limit. At large separations r_{21} of the spheres under consideration, when the phase shift factor decreases more rapidly than the multipole susceptibilities, we replace the susceptibilities by their zero frequency limit. Only the low frequency modes, whose phase shift is still negligibly small, contribute to the retarded limit.

The reduction of the general dispersion function $G(\omega, \mu)$ given by the determinant (6.6) to the nonretarded limit is achieved by using

$$\Delta_{ss}(m, j) \Rightarrow (2^{2m} m!/(2m)!)^2 (K R_j)^{2m+1}$$
(6.11)

$$\begin{cases} (m+1)(\mu_j - \mu)/[m\mu_j + (m+1)\mu] \\ (m+1)(\varepsilon_j - \varepsilon)/[m\varepsilon_j + (m+1)\varepsilon] \end{cases} \quad \text{for} \quad s = \begin{cases} 1 \\ 2 \end{cases}$$

and

$$V_{mn}^{\mu}(Kr_{21}) \Rightarrow -i(Kr_{21})^{-(m+n+1)}[(m+n)!/(m-\mu)!(n-\mu)!]$$

$$\cdot [(2m)!/2^m(m-1)!] \, [(2n)!/2^n(n-1)!]$$
(6.12)

$$W_{mn}^{\mu}(Kr_{21}) \Rightarrow 0 \, .$$

$G(\omega, \mu)$ splits into an electric and a magnetic subdeterminant with the former being equal to the nonretarded secular determinant found in Section 4.3.

To integrate the energy formula (6.8) in the retarded limit, we note from Eq. (6.11) that the zero frequency limit of $\Delta_{ss}(m, j)$ is approached proportional to K^{2m+1}. Generally introducing the reduced variable $\zeta = Kr_{21}$, we find that the multipole interaction term m, n is proportional to $r_{21}^{-(2m+2n+3)}$. The lowest order term, the dipole-dipole interaction term $m = n = 1$, is proportional to r_{21}^{-7}.

The recovery of all previous findings, in spite of numerous simplifications, clearly demonstrates the strength and generality of the procedure used.

6.2. Cylinders

The calculation of dispersion energies between cylinders is fully analogous to that in the presence of spheres. We are interested in the dispersion function $G(\omega, k)$ on the imaginary frequency axis where the radial wave number k is also imaginary according to the relationship (5.17). $J_m(k\varrho)$ as well as $Y_m(k\varrho)$ increase exponentially with increasing radius ϱ, i.e., the boundary condition (5.47) caused by the cavity with radius r_S turns into

$$\lim_{r_S \to \infty} a_{2s}(m, j)/a_{1s}(m, j) = -\lim_{r_S \to \infty} J_m(k r_S)/Y_m(k r_S)$$

$$= \pm i \quad \text{for} \quad \text{Im}(k) \gtrless 0 \, .$$
(6.13)

Substituting Eq. (6.13) into the potential (5.46) we obtain in a similar manner to the case of spheres

$$a_{1s}(m,j)\,J_m(k\varrho) + a_{2s}(m,j)\,Y_m(k\varrho)$$

$$= a_{1s}(m,j) \begin{cases} H_m^{(1)}(k\varrho) \\ H_m^{(2)}(k\varrho) \end{cases} \quad \text{for} \quad \text{Im}(k) \gtrless 0. \tag{6.14}$$

The effect of the cavity is that the external potential $Z(r)$ is composed of Bessel functions of the third kind, which decrease exponentially with increasing radius ϱ for imaginary frequencies. Inserting Eq. (6.13) into boundary conditions (5.48) relating to the cylinder surfaces, we note that they are not yet in diagonal form for infinite separation r_{21}. There is a coupling of the ingoing electric and magnetic modes at each cylinder j, which suggests that diagonalized linear combinations should be used from the start. This is especially the case in view of the fact that the dispersion function $G(\omega, k)$ entering in the calculation of ΔE_{12} equals the ratio of secular determinants resulting from Eq. (5.48) for finite and infinite separation r_{21}. Retention of a nondiagonalized determinant in the denominator would strongly impede the calculation of the dispersion energy.

In order to diagonalize expression (5.48) we note that on satisfying the Helmholtz equation in cylindrical coordinates according to (5.16), we are not limited to describing the outgoing modes in terms of Bessel functions $Y_m(k\varrho)$ of the second kind. We may substitute any linear combination of $J_m(k\varrho)$ and $Y_m(k\varrho)$ if we substitute accordingly in boundary conditions (5.27)–(5.30). By substituting in particular the Bessel functions $H_m^{(1)}(k\varrho)$ or $H_m^{(2)}(k\varrho)$ of the third kind, we note from Eq. (5.47) that the coefficients $a_{1s}(m,j)$ vanish at imaginary frequencies, so that we are left with the diagonalized form of boundary conditions (5.48).

Hence, by summing over all translation wave numbers k_z we find

$$\Delta E_{12} = (\hbar/4\pi i) \int_{-i\infty}^{+i\infty} d\omega \coth(\hbar\omega/2kT)\,(L/2\pi) \int_{-\infty}^{+\infty} dk_z \ln G(\omega, k) \tag{6.15}$$

where

$$G(\omega, k) = \det \begin{pmatrix} 1 & 0 & \Delta_{11}(m, 1)\,H_{m+n}^{(1)} & \Delta_{21}(m, 1)\,H_{m+n}^{(1)} \\ 0 & 1 & \Delta_{12}(m, 1)\,H_{m+n}^{(1)} & \Delta_{22}(m, 1)\,H_{m+n}^{(1)} \\ \Delta_{11}(n, 2)\,H_{m+n}^{(1)} & \Delta_{21}(n, 2)\,H_{m+n}^{(1)} & 1 & 0 \\ \Delta_{12}(n, 2)\,H_{m+n}^{(1)} & \Delta_{22}(n, 2)\,H_{m+n}^{(1)} & 0 & 1 \end{pmatrix}. \tag{6.16}$$

The integration over k_z in Eq. (6.15) can be replaced by that over k according to relationship (5.17). Changing the sign of k_z does not affect

$G(\omega, k)$, it inverts the electric modes relative to the magnetic modes. By using Eq. (5.50) we may split $G(\omega, k)$ into two subdeterminants with the indices m, n of their elements merely running over positive values.

By considering the nonretarded limit of Eqs. (6.15), (6.16) we readily reobtain Eqs. (4.57)–(4.59). The modified Bessel function $K_m(\zeta)$ of the second kind equals the Bessel function $H_m^{(1)}(i\,\zeta)$ of the third kind except for a constant factor, which cancels in the product $\Delta_{st}(m, 1)\,\Delta_{ts}(n, 2)$ $(H_{m+n}^{(1)}(i\zeta))^2$. In addition to the electrostatic interaction due to the electric multipole susceptibility $\Delta_{22}(m, j)$, we find a magnetostatic interaction due to the magnetic multipole susceptibility $\Delta_{11}(m, j)$.

By expanding $\ln G(\omega, k)$ with respect to its off-diagonal elements up to quadratic terms we obtain

$$\Delta E_{12} = -\,(\hbar/4\pi i) \oint_{-i\infty}^{+i\infty} d\omega \coth(\hbar\omega/2kT)\,(L/2\pi)\int_{-\infty}^{+\infty} dk_z \sum_{s,t=1}^{2} \sum_{m,n=-\infty}^{+\infty}$$

$$\cdot\,\Delta_{st}(m, 1)\,H_{m+n}^{(1)}(k\,r_{21})\,\Delta_{ts}(n, 2)\,H_{n+m}^{(1)}(k\,r_{21})\,. \tag{6.17}$$

There are electric, magnetic and mixed contributions to the total dispersion energy, with the sign of the latter depending on the order of the multipoles m and n under consideration.

The convergence of all multipole contributions m, n in Eq. (6.17) is guaranteed by the decrease of the multipole susceptibilities and of the coupling parameters $H_{m+n}^{(1)}(k\,r_{21})$ with increasing imaginary frequency and increasing k_z. From Eqs. (9.6.4) and (9.7.2) in Ref. [1] we find

$$i^{m+1}\,H_m^{(1)}(i\,\zeta) = (2/\pi\zeta)^{1/2} \exp(-\,\zeta)\,[1 + (2m+1)(2m-1)/8\zeta + \cdots] \tag{6.18}$$

i.e., $(H_{m+n}^{(1)}(k\,r_{21})^2$ is proportional to $\exp(-\,2\zeta)/\zeta$ with increasing ζ, $i\zeta = k\,r_{21}$. Whether this phase shift factor or the multipole susceptibilities decrease more rapidly with increasing imaginary frequency depends on the separation r_{21} of the cylinders under investigation. The nonretarded limit has been extensively discussed in Sections 4.4 and 4.6. In the retarded limit, when the phase shift of the interacting modes is large, we may replace the multipole susceptibilities $\Delta_{st}(m, j)$ by their limiting value for zero frequency, which is approached proportional to k^{2m}. Then replacing the integrals over ω and k_z by an integral over k according to Eq. (5.17) we find the multipole interaction term m, n in Eq. (5.70) to vary with $r_{21}^{-(2m+2n+2)}$. The retarded contributions to the dispersion energy exhibit an additional factor r_{21}^{-1} compared with the nonretarded result according to Eq. (4.62).

Comparison of Eqs. (6.8), (6.9) with Eqs. (6.15), (6.16) shows the close agreement of the final results in the case of spheres and cylinders. It is possible in both cases to represent the coupling parameters by modified

Bessel functions $K_m(\zeta)$ of the second kind. In the case of spheres, m is half an integer and ζ equals $K r_{21}$ whereas in the case of cylinders m is an integer and ζ equals $k r_{21}$.

6.3. Half-Spaces

Finally, let us consider the retarded dispersion energy between half-spaces. Since the cavity now cuts off the internal rather than the external modes, it is the multipole susceptibilities $\Delta_{ss}(k,j)$ which are affected by taking the limit $x_S \to \infty$. When ω is imaginary we find from Eq. (5.19) that k, as well as k_1 and k_2, all become imaginary. Then, using

$$\lim_{x_S \to \infty} \coth k_j(x_j - x_S) = \mp i \quad \text{for} \quad \text{Im}(k_j) \gtrless 0 \tag{6.19}$$

we obtain

$$\Delta E_{12} = (\hbar/4\pi i) \int_{-i\infty}^{+i\infty} d\omega \coth(\hbar\omega/2kT)(L/2\pi)^2 \int_{-\infty}^{+\infty} dk_y \int_{-\infty}^{+\infty} dk_z$$
$$\cdot \sum_s \ln G_s(\omega, k) \tag{6.20}$$

where

$$G_1(\omega, k) = 1 - \exp(2ik x_{21}) \frac{k_1 \mu_1^{-1} - k\mu^{-1}}{k_1 \mu_1^{-1} + k\mu^{-1}} \frac{k_2 \mu_2^{-1} - k\mu^{-1}}{k_2 \mu_2^{-1} + k\mu^{-1}} \tag{6.21}$$

$$G_2(\omega, k) = 1 - \exp(2ik x_{21}) \frac{k_1 \varepsilon_1^{-1} - k\varepsilon^{-1}}{k_1 \varepsilon_1^{-1} + k\varepsilon^{-1}} \frac{k_2 \varepsilon_2^{-1} - k\varepsilon^{-1}}{k_2 \varepsilon_2^{-1} + k\varepsilon^{-1}} . \tag{6.22}$$

On comparing this result with the nonretarded findings in Section 4.5 we find i) an additional magnetic contribution, ii) a slight dependence of $\Delta_{ss}(k,j)$ on k, iii) k depends on the frequency ω.

It is the latter property which causes the main difference between the retarded and the nonretarded limit. At small separations r_{21}, when the multipole susceptibilities $\Delta_{ss}(m,j)$ decrease more rapidly with increasing imaginary frequency than the phase shift factor $\exp(-2\zeta)$, $i\zeta = kx_{21}$, we may treat the ω integral and the k_y, k_z integrals independently, yielding Eq. (4.76). At large separations x_{21}, when the phase shift factor decreases more rapidly than the multipole susceptibilities $\Delta_{ss}(k,j)$, we replace the latter by their zero frequency limit. Then, by joining the ω and the k_y, k_z integrals to give $\int d(ik) \int dk_y \int dk_z = 4\pi \int d(ik)(ik)^2$ according to relationship (5.19) we find

$$\Delta E_{12} = \hbar c(\varepsilon\mu)^{-\frac{1}{2}}(L/2\pi)^2 x_{21}^{-3} \int_0^\infty d\zeta \, \zeta^2 \sum_s \ln(1 - e^{-2\zeta} \Delta_{ss}(0,1) \Delta_{ss}(0,2)). \tag{6.23}$$

The retarded dispersion energy between half-spaces is proportional to the inverse cube of the separation x_{21}. As in the case of spheres or cylinders, we find an additional power x_{21}^{-1} compared with the nonretarded result (4.76).

The retarded dispersion energy between half-spaces was first considered by Lifshitz in 1955 [28]. We outlined his procedure in Section 5.1.

6.4. Summary

We have now developed and applied a macroscopic theory which is formally complete. We inquired about the free energy gain of the electromagnetic field in the presence of two particles 1 and 2.

The first step is to solve the Helmholtz equation in the presence of a single particle j. Explicit solutions of the Helmholtz equation are known in rectangular coordinates, in circular, elliptic and parabolic cylinder coordinates, and in spherical and conical coordinates, see Ref. [4]. Generally we find two times two types of solutions which represent ingoing and outgoing radiation and electric and magnetic modes, respectively. By requiring normalizability of all solutions in the interior of particle j and satisfying the continuity conditions for the electric and the magnetic field across the surface, we obtain a linear set of equations between the amplitudes of the external modes. The amplitudes of the outgoing modes depend linearly on those of the ingoing modes, i.e., the respective coefficients can be interpreted as multipole susceptibilities of particle j. Normalizability in the exterior is then achieved by introducing a large perfectly reflecting cavity.

The second step is to solve the Helmholtz equation in the presence of the two particles 1 and 2. We use linear combinations of the solutions found in the presence of the single particles. The outgoing modes of particle 2 present the ingoing modes for particle 1, and vice versa. By transposing all modes to a single particle and there satisfying the linear relations between the amplitudes of the ingoing and outgoing modes, we obtain the secular determinant (dispersion function) for the eigenfrequencies of the normalizable solutions.

To find the total energy gain of all modes relative to the case of infinite separations of particles 1 and 2 we use the theorem of residues. The eigenfrequencies of the normalizable solutions, being the zeros of the respective secular determinant, are the poles of the logarithmic derivative of this dispersion function. We may replace the summation over the free energy gain of all modes by a contour integral around the logarithmic derivative of the dispersion function. The contour of integration can be shifted to the imaginary frequency axis, but has to by-pass the poles of the Bose distribution $\coth(\hbar\omega/2kT)$ from the right.

The change to imaginary frequencies entails mathematical and physical simplifications as well. The ingoing and outgoing modes split into exponentially increasing and decreasing components. The cavity requires that only the decreasing component is left. The size of the cavity can be increased towards infinity. We are then interested in those solutions of the Helmholtz equation where the ingoing and outgoing radiation balance each other in the upper or in the lower half-frequency plane. Only those modes are permitted which are normalizable in infinite space in the presence of stimulation or damping, i.e. under the action of the random energy exchange with the rest of modes. We may neglect the modes exchanging energy with the exterior. They do not explicitly contribute to the attraction between particles 1 and 2.

It is this selective interaction which justifies the effort involved in constructing and transposing optimally adapted potentials in spherical, cylindrical, or rectangular coordinates. It is now permissible to limit the generally infinite secular determinant to a finite number of elements or to cut off the perturbation expansion after a finite number of terms. Each multipole considered increases the order of the respective secular determinant $G(\omega, k)$ by four: The multipole may be situated at particle 1 or 2 and may be electric or magnetic. Treating dipole interactions yields a secular determinant of order four, the inclusion of quadrupole interactions causes a secular determinant of order eight.

The main difference between the retarded and the nonretarded findings arises because of the question as to whether it is the multipole susceptibility of the particles or the phase shift and strength of the fields which decreases more rapidly with increasing imaginary frequency and wave number. The final energy expression results from a frequency integral along the imaginary axis and from n wave number integrals, with $n = 0$ for spheres, $n = 1$ for cylinders, and $n = 2$ for half-spaces. If the multipole susceptibilities of the particles decrease more rapidly with increasing imaginary frequency than the phase shift and strength of the fields, we may split the total integral into a frequency integral over the multipole susceptibilities and into n wave number integrals over the field amplitudes. On the other hand, if the multipole susceptibilities decrease only slowly, we may take their zero frequency limit and in that case are left with $n + 1$ wave number integrals over the field amplitudes. Introducing the reduced variable $i\zeta = kr_{21}$, we generally obtain one additional power r_{21}^{-1} compared with the nonretarded limit, where only n wave number integrals are to be considered.

On principle, we may consider solving the problem represented above solely on the basis of plane waves. The planar potentials (5.18) arising in rectangular coordinates represent a complete set of transverse solutions of the Helmholtz equation, and thus render the correct secular determinant and the complete set of eigenfrequencies of the system under investigation. However, in that case it is not possible to limit the secular determinant to a finite number of elements or to use finite perturbation theory. Before doing so we have to return to linear combinations of the planar modes considered, which are optimally adapted to the particles under investigation. We therefore end up where we started.

Nevertheless, there is also a possibility of finding the dispersion energy between particles 1 and 2 from a perturbation theory with respect to planar modes: It is a general principle of perturbation theory that the total shift of eigenvalues equals zero. The trace of the secular matrix is preserved on any linear transformation of the basis functions. This means in the case of the Helmholtz equation that the total shift of the squared frequencies vanishes. This principle holds for the shift of the frequencies themselves if the perturbation considered maintains the overall state density, which applies to the interaction of the modes localized around the particles with the nonlocalized free modes, i.e. to the difference between the nonretarded and the retarded interaction. If the nonretarded interaction between particles 1 and 2 is already known, we may find the retarded interaction from perturbation theory with respect to the free modes. This fact was used by Casimir and Polder in their investigations into the retarded dispersion energy between atoms [26]. The general procedure presented here shows that, by shifting the contour of integration to the imaginary axis without regard to the necessary crossing of poles, it is possible to reobtain the eigenfrequencies of the localized modes, which are zeros of the infinite secular determinant but which are missing in the perturbation expansion. This principle was used by Langbein in 1970 in investigations into the retarded dispersion energy between spheres on the basis of perturbation theory [33]. This indirect method appears to be most powerful whenever the shape or composition of the particles considered makes it impossible to find well adapted modes.

In contrast to the nonretarded treatment, which is solely based on the electrostatic interaction potential $V(r)$, we equate the electric potential to zero in the retarded case. The electric interaction is completely covered by the vector potential $A(r)$. The transverse modes under investigation enable the Lorentz gauge to be used for vanishing electric potential. This concept turns out especially useful with respect to the Schrödinger formalism presented in the next Chapter. We may ascribe the retarded interaction between electrons located at different particles solely to the vector potential $A(r)$. In addition to providing the proper multipole susceptibilities, quantum theory still has to answer the question regarding statistics. Are the localized modes which are strictly coupled to molecular electron transitions, still Bosons?

6.5. Lattices

The procedure presented here is obviously not restricted to pairs of macroscopic particles. We may consider arbitrary arrays in a similar manner. An immediate effect is a lowering of the symmetry. We only can classify the allowed electromagnetic modes according to their rotational behavior in the presence of a linear array of spheres. As soon as the rotational invariance gets violated we have to construct the allowed modes from linear combinations of arbitrarily oriented multipoles. We have to take into account three, rather than one, dipoles, five, rather than one, quadrupoles, and so on. This increases considerably the order of the resulting secular problem. However, a new reduction of the number of interacting modes is possible if the array considered exhibits some further symmetry. Any allowed mode remains an allowed mode on application of any possible symmetry operation. If the array under consideration is invariant against inversion, as is true in the presence of two identical spheres or cylinders, the allowed modes are even or odd with respect to inversion. If we consider a tetrahedral array, we find modes which are invariant and others which pick up a phase factor $\exp(2n\pi i/3)$ on rotation by $2\pi/3$. When investigating a periodic lattice, we conclude, from the equivalence of all lattice sites, that there are allowed modes gaining a phase factor $\exp(i\mathbf{q} \cdot \mathbf{r}_j)$ on translation by any lattice vector $\mathbf{r}_j$. This fact is known as Floquet's theorem in mathematics, and as Bloch's theorem in solid state physics. The use of Bloch's theorem reduces the order of the secular determinants arising in a periodic lattice to the number of multipoles considered at the individual particles.

Constructing exact eigenvectors of the Helmholtz equation in a periodic array of spheres is of great value for treating the lattice energy of the inert gas crystals. It is in these lattices where the usual perturbation approach to van der Waals attraction appears to be most unsatisfactory. None of the multiplet or higher order contributions favor the face-centered cubic structure sufficiently relative to the hexagonal close-packed structure to prevent the next higher order terms from reversing the result.

Let us outline the procedure. The proper linear combinations of spherical modes in a lattice are the Bloch sums

$$Z(r) = \sum_j \exp(i\mathbf{q} \cdot \mathbf{r}_j) \sum_{s=1}^{2} (K^{-1}\mathbf{\nabla} \times)^s (\mathbf{r} - \mathbf{r}_j) \sum_{m,\mu}$$

$$\cdot [a_{1s}(m,\mu) j_m(K|\mathbf{r} - \mathbf{r}_j|) + a_{2s}(m,\mu) y_m(K|\mathbf{r} - \mathbf{r}_j|)] \tag{6.24}$$

$$\cdot P_m^\mu(\cos \vartheta_j) \exp(i\mu\varphi_j)/(m+\mu)!$$

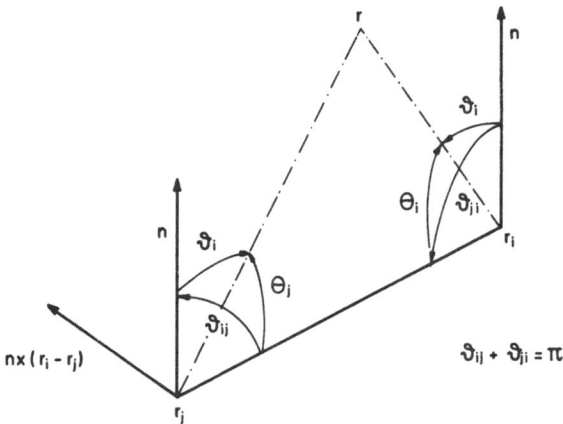

Fig. 26. Rotation of angles

The amplitudes $a_{1s}(m, \mu)$, $a_{2s}(m, \mu)$ of the potential at the different lattice sites r_j are interrelated by the phase factor $\exp(i q \cdot r_j)$. Their second argument is no longer the particle number j but the rotation wave number μ. Though we know that we are finally interested only in exponentially decreasing external potentials, we use $j_m(Kr)$ as well as $y_m(Kr)$ in order to be able to correctly satisfy the continuity conditions at all lattice sites. All spherical coordinates used in Eq. (6.24) are oriented in a standard direction n. The use of inversely oriented spherical coordinates is possible only in the presence of pairs of spheres and would contradict the use of Bloch's theorem.

To satisfy the continuity of the field across the surface of sphere i we have to transpose all modes to lattice site r_i. This implies that we have to rotate the coordinates at lattice site r_j to the connecting line $r_i - r_j$ to site r_i, and transpose all modes from site j to site i by means of addition theorem (5.36), and rotate the coordinates back to the standard direction n. The rotation of coordinates couples multipoles of the same type but different orientation. Using the notation shown in Fig. 26 we have

$$P_m^\mu(\cos \vartheta_j) \exp[i\mu(\varphi_j + \tfrac{1}{2}\pi)]/(m + \mu)!$$

$$= \sum_{\nu = -m}^{+m} C(m, \mu, \nu, \vartheta_{ij}) P_m^\nu(\cos \theta_j) \exp[i\nu(\phi_j - \tfrac{1}{2}\pi)]/(m + \nu)! \tag{6.25}$$

where the azimuth angles φ_j and ϕ_j in systems n and $r_i - r_j$ are measured relative to their axis of intersection $n \times (r_i - r_j)$, and where $C(m, \mu, \nu, \vartheta)$

may be represented by a Jacobi polynomial of argument $\sin^2 \vartheta/2$,

$$C(m, \mu, v, \vartheta) = (\cos\tfrac{1}{2}\vartheta)^{\mu+v} \frac{(m-v)!}{(m+\mu)!} \sum_\lambda \frac{(-1)^\lambda (2m-\lambda)!}{(m-\mu-\lambda)!\,(m-v-\lambda)!\,\lambda!}$$

$$\cdot (\sin\tfrac{1}{2}\theta)^{2m-2\lambda-\mu-v}\,. \tag{6.26}$$

By using rotation theorem (6.25) at site j, transposing all modes to site i by means of Eq. (5.36), using rotation theorem (5.38) at site i, and satisfying continuity conditions (5.23) across the surface of sphere i, we obtain

$$\Delta_{ss}(m, i) \Big\{ a_{1\,s}(m, \mu) + (2m+1) \sum_{j \neq i} \exp(i\boldsymbol{q}\cdot\boldsymbol{r}_j) \sum_{n, v, \lambda} \exp(iv\varphi_{ji}) C(n, v, -\lambda, \vartheta_{ij})$$

$$\cdot (-1)^{n-v} [(m-\lambda)!/(n+\lambda)!]\, [a_s(n, v)\, V^\lambda_{nm}(Kr_{ji}) + a_t(n, v)\, W^\lambda_{nm}(Kr_{ji})]$$

$$\cdot C(m, -\lambda, \mu, -\vartheta_{ij}) \exp(-i\mu\varphi_{ji}) \Big\} + a_{2\,s}(m, \mu) = 0\,. \tag{6.27}$$

ϑ_{ji}, φ_{ji} are the polar angle and the azimuth of $\boldsymbol{r}_{ji} = \boldsymbol{r}_j - \boldsymbol{r}_i$ with respect to the standard system $\boldsymbol{n}$ and t equals 2 for $s = 1$, and vice versa.

The various sums and factors in Eq. (6.27) may clearly be attributed to the sequence rotation – transposition – rotation. As a result of using Bloch's theorem, we find that by satisfying the continuity conditions at site i, those at all other sites are automatically satisfied as well. The secular system (6.27) depends on the relative position $\boldsymbol{r}_{ji}$ of sites j and i only. The ingoing modes at site i are linearly composed of the outgoing modes at site j. Eq. (6.27) contains the lattice sums over outgoing multi-poles with arbitrary orientation at all sites j. On changing the sign of $\boldsymbol{q}$ and counting lattice sites $\boldsymbol{r}_j$ relative to $\boldsymbol{r}_i$, we find an inverted coupling between electric and magnetic modes which, however, does not change the spectrum of eigenfrequencies.

Equations (6.27) permit an exact calculation of the allowed frequency spectrum in a lattice of spheres from that of a single sphere. Each single line splits into a band. Modes belonging to the same band are distinguished by the wave number $\boldsymbol{q}$.

To obtain the lattice energy we use the state density integration described in Section 3.4. Turning to an infinite cavity radius r_S according to Eq. (6.1), we find

$$\Delta E = (\hbar/4\pi i) \oint_{-i\infty}^{+i\infty} d\omega \coth(\hbar\omega/2kT)(2\pi)^{-3} \int d\boldsymbol{q} \ln G(\omega, \boldsymbol{q}) \tag{6.28}$$

where $G(\omega, \boldsymbol{q})$ is the ratio of secular determinants resulting from Eq. (6.27) for finite and infinite lattice spacing, and the integral over $\boldsymbol{q}$ covers one reciprocal lattice cell. If only dipole interactions $m = 1$; $\mu = -1, 0, +1$

are considered, $G(\omega, q)$ turns out to be a determinant of the sixth order. If quadrupole interactions are included, we are left with a determinant of the sixteenth order. All diagonal elements of $G(\omega, q)$ equal 1, all off-diagonal elements decrease exponentially with increasing imaginary frequency, i.e., along the contour of integration in Eq. (6.28).

7. Schrödinger Formalism

7.1. Perturbation Theory

In the preceding chapter we considered in full detail the electrodynamic aspect of van der Waals attraction. Solving for the eigenvectors of the vector Helmholtz equation in the presence of particles 1 and 2, we found the main contribution to the dispersion energy caused by those eigenvectors which arise from a balance between ingoing and outgoing radiation and are localized in the vicinity of particles 1 and 2. We accounted for quantum theory merely by assuming all eigenvectors to obey Bose statistics.

The electrodynamic approach stresses the photon, but ignores the electron aspect of van der Waals attraction. We take account of the electrons only globally by assigning polarizabilities to the interacting molecules or permeabilities to the interacting particles. This rating is inverted if we adopt the viewpoint of quantum mechanics and calculate the correlation energy of electron orbitals from the Schrödinger equation. Now the photons are strongly undervalued. They merely make up for the electronic interaction potential. This interaction potential may be taken to be electrostatic as in the usual Schrödinger formalism, or may be identified with the electrodynamic four potential.

Let us first follow London's original procedure, that is, let us solve the two particle Schrödinger equation by means of perturbation theory. We consider two particles 1 and 2 at positions r_1 and r_2, which exhibit the one-electron Hartree-Fock or pseudopotentials $U_1(r)$ and $U_2(r)$, as shown in Fig. 27. We denote the one-electron orbitals localized at particle 1 by $|i\rangle, |k\rangle, |m\rangle, \ldots$, those localized at particle 2 by $|j\rangle, |l\rangle$,

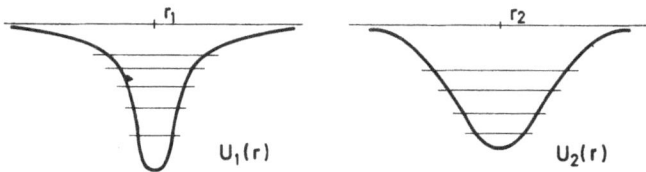

Fig. 27. One-electron potentials

$|n\rangle, \dots$. Then, in the case of infinite separation $r_{21} = r_2 - r_1$, we may start with the decoupled Schrödinger equations

$$\{T(r) + U_1(r) - E_i\} |i\rangle = 0 \tag{7.1}$$

$$\{T(r) + U_2(r) - E_j\} |j\rangle = 0 \tag{7.2}$$

where

$$T(r) = -(\hbar^2/2m) \, \nabla \cdot \nabla \tag{7.3}$$

is the operator of kinetic energy.

We are interested in studying the correlation energy of the two-electron orbitals of particles 1 and 2. Thus, turning to finite separations $r_{21} = r_2 - r_1$ we consider the Schrödinger equation

$$\{T(r_i) + T(r_j) + U_1(r_i) + U_2(r_j) + W(r_i, r_j) - E\} \, \psi(r_i, r_j) = 0 \tag{7.4}$$

with the coordinates r_i and r_j designating electrons at particles 1 and 2, respectively.

The interaction potential $W(r_i, r_j)$ is built up from the exchange terms $U_1(r_j)$ and $U_2(r_i)$ such that electron i sees particle 2 and electron j sees particle 1, and from the Coulomb repulsion of the electrons $V_{el}(r_i, r_j)$ plus that of the remaining ions $V_{ion}(r_1, r_2)$, i.e.

$$W(r_i, r_j) = U_1(r_j) + U_2(r_i) + V_{el}(r_i, r_j) + V_{ion}(r_1, r_2) \, . \tag{7.5}$$

In the following we assume the separation r_{21} to be large enough to exclude overlap of electron orbitals localized at different particles. In this case it is permissible to omit all exchange terms in the two-electron orbitals $\psi(r_i, r_j)$ and to write

$$\psi(r_i, r_j) = \sum_{i,j} a_{ij} |i\rangle \, |j\rangle \, . \tag{7.6}$$

The two-electron orbital $\psi(r_i, r_j)$ is a linear combination of products of one-electron orbitals localized at different particles. Applying second order perturbation theory to the Schrödinger equation (7.4), the energy of the orbital specified by (7.6) is found to be

$$E - E_i - E_j = \langle ij| W |ij\rangle + \sum_k \frac{\langle ij| W |kj\rangle \langle kj| W |ij\rangle}{E_i - E_k}$$

$$+ \sum_l \frac{\langle ij| W |il\rangle \langle il| W |ij\rangle}{E_j - E_l} + \sum_{k,l} \frac{\langle ij| W |kl\rangle \langle kl| W |ij\rangle}{E_i - E_k + E_j - E_l} + \cdots \tag{7.7}$$

with k, l running over all empty orbitals of particles 1 and 2. The total interaction energy of particles 1 and 2 results from Eq. (7.7) by summation over all occupied electron orbitals $|i\rangle$ and $|j\rangle$. Introducing occupation numbers $f_i, f_j = 0, 1$, we define the free energy of interaction of particles 1 and 2 by

$$\Delta E_{12} = \sum_{i,j} f_i f_j (E - E_i - E_j) \tag{7.8}$$

yielding

$$\Delta E_{12} = \Delta E_{\text{or}} + \Delta E_{\text{ind}} + \Delta E_{\text{dis}} \tag{7.9}$$

where

$$\Delta E_{\text{or}} = \sum_{i,j} f_i f_j \langle ij | W | ij \rangle \tag{7.10}$$

$$\Delta E_{\text{ind}} = \sum_{i,j,k} f_i (1 - f_k) f_j \frac{|\langle ij | W | kj \rangle|^2}{E_i - E_k} + \sum_{i,j,l} f_i f_j (1 - f_l) \frac{|\langle ij | W | il \rangle|^2}{E_j - E_l} \tag{7.11}$$

$$\Delta E_{\text{dis}} = \sum_{i,k} f_i (1 - f_k) \sum_{j,l} f_j (1 - f_l) \frac{|\langle ij | W | kl \rangle|^2}{E_i - E_k + E_j - E_l}. \tag{7.12}$$

The three terms ΔE_{or}, ΔE_{ind} and ΔE_{dis} correspond to and generalize the orientation energy, the induction energy, and the dispersion energy discussed in Section 1.2. Expanding the interaction potential $W(r_i, r_j)$ into a Taylor series in $r_i - r_1$ and $r_j - r_2$ up to terms of the second order and using the fact that no external potential may remain if the position of an electron coincides with the center of the respective potential, we obtain

$$\Delta E_{\text{or}} = \sum_i f_i \langle i | r - r_1 | i \rangle \cdot V_1 V_2 V_{\text{el}} \cdot \sum_j f_j \langle j | r - r_2 | j \rangle \tag{7.13}$$

$$\Delta E_{\text{ind}} = \sum_{i,k} f_i (1 - f_k) \sum_j f_j \frac{|\langle i | r - r_1 | k \rangle \cdot V_1 V_2 V_{\text{el}} \cdot \langle j | r - r_2 | j \rangle|^2}{E_i - E_k}$$
$$+ \sum_i f_i \sum_{j,l} f_j (1 - f_l) \frac{|\langle i | r - r_1 | i \rangle \cdot V_1 V_2 V_{\text{el}} \cdot \langle j | r - r_2 | l \rangle|^2}{E_j - E_l} \tag{7.14}$$

$$\Delta E_{\text{dis}} = \sum_{i,k} f_i (1 - f_k) \sum_{j,l} f_j (1 - f_l) \frac{|\langle i | r - r_1 | k \rangle \cdot V_1 V_2 V_{\text{el}} \cdot \langle j | r - r_2 | l \rangle|^2}{E_i - E_k + E_j - E_l}. \tag{7.15}$$

The first order term ΔE_{or} is proportional to the permanent dipole moments

$$p_1 = \sum_i f_i \langle i | r - r_1 | i \rangle; \quad p_2 = \sum_j f_j \langle j | r - r_2 | j \rangle \tag{7.16}$$

of particles 1 and 2,

$$\Delta E_{\text{or}} = p_1 \cdot V_1 V_2 V_{\text{el}}(r_1, r_2) \cdot p_2. \tag{7.17}$$

This is just the classical expression (1.2). ΔE_{or} depends on the mutual orientation of particles 1 and 2, i.e., the permanent dipoles p_1 and p_2 tend to align.

In order to simplify the second-order terms ΔE_{ind} and ΔE_{dis} in a similar manner, we introduce the dipole susceptibility tensors of particles 1 and 2

$$\mathbf{X}_1(\omega) = \sum_{i,k} f_i(1 - f_k) \langle i|r - r_1|k\rangle \langle k|r - r_1|i\rangle$$
$$\cdot [(E_k - E_i + \hbar\omega)^{-1} + (E_k - E_i - \hbar\omega)^{-1}] \tag{7.18}$$

$$\mathbf{X}_2(\omega) = \sum_{j,l} f_j(1 - f_l) \langle j|r - r_2|l\rangle \langle l|r - r_2|j\rangle$$
$$\cdot [(E_l - E_j + \hbar\omega)^{-1} + (E_l - E_j - \hbar\omega)^{-1}] \tag{7.19}$$

yielding

$$\Delta E_{ind} = -\tfrac{1}{2} p_2 \cdot \boldsymbol{V}_1 \boldsymbol{V}_2 V_{el} \cdot \mathbf{X}_1(0) \cdot \boldsymbol{V}_1 \boldsymbol{V}_2 V_{el} \cdot p_2$$
$$-\tfrac{1}{2} p_1 \cdot \boldsymbol{V}_1 \boldsymbol{V}_2 V_{el} \cdot \mathbf{X}_2(0) \cdot \boldsymbol{V}_1 \boldsymbol{V}_2 V_{el} \cdot p_1 . \tag{7.20}$$

We reobtain the classical induction energy (1.5). The permanent dipole p_1 at particle 1 polarizes particle 2, the induced field lowers the energy of dipole p_1, and vice versa.

By comparing ΔE_{dis} according to Eq. (7.15) with expressions (7.18), (7.19), we note that ΔE_{dis} contains $\mathbf{X}_1(\omega)$ at the pole $\hbar\omega = E_l - E_j$ of $\mathbf{X}_2(\omega)$, and $\mathbf{X}_2(\omega)$ at the pole $\hbar\omega = E_k - E_i$ of $\mathbf{X}_1(\omega)$. Again, a complex contour integration seems very appropriate. Introducing the identity

$$(2\pi i)^{-1} \int_{-i\infty}^{+i\infty} d(\hbar\omega) [(E_k - E_i + \hbar\omega)^{-1} + (E_k - E_i - \hbar\omega)^{-1}]$$
$$\cdot [(E_l - E_j + \hbar\omega)^{-1} + (E_l - E_j - \hbar\omega)^{-1}] = 2(E_k - E_i + E_l - E_j)^{-1} \tag{7.21}$$

we find

$$\Delta E_{dis} = -(\hbar/4\pi i) \int_{-i\infty}^{+i\infty} d\omega \, \mathrm{tr}\{\mathbf{X}_1(\omega) \cdot \boldsymbol{V}_1 \boldsymbol{V}_2 V_{el} \cdot \mathbf{X}_2(\omega) \cdot \boldsymbol{V}_1 \boldsymbol{V}_2 V_{el}\}. \tag{7.22}$$

We reobtain the energy expression derived from the fluctuation approach and from the oscillator model in Chapter 3. Higher order contributions to ΔE_{dis} are found by applying higher order perturbation theory analogous to the procedure given in Section 3.3.

Although the Schrödinger equation does not suggest anything about the origin of the electronic interaction potential, we obtain an energy expression which may be understood to result from a photon exchange between the particles under consideration. The frequency integration along the imaginary frequency axis, which, as was the case in the oscillator model, seems artificial at first sight, arises when looking for the eigen-frequencies of the coupled system of particles plus electromagnetic fields and using the state density integration described in Section 3.4.

7.2. Reaction Potentials

The main difference between the Schrödinger formalism and the discussed semiclassical approaches lies in statistics. The electrons obey Fermi statistics, i.e., the occupation numbers f_i, f_j entering the present calculations on average follow the Fermi distribution. To use the identity (7.21), we have to require that the energy E_k of the empty states $|k\rangle$ is larger than the energy E_i of the occupied states $|i\rangle$. With increasing temperature, when this condition gets violated, a modified integration has to be used. It seems appropriate to choose a contour of integration enclosing the full real axis and a weight factor differing by one in the right-hand and the left-hand half-plane. However, without going into the details of the interaction, there is no reason to prefer the Bose factor $\frac{1}{2}\coth(\hbar\omega/2kT)$ to the Fermi factor $\frac{1}{2}\tanh(\hbar\omega/2kT)$. Hence, let us once more stress the viewpoint of oscillating interaction potentials. Let us consider the reaction of particle 1, whose one-electron orbitals satisfy the Schrödinger equation (7.1), to an oscillating external potential $\delta U(r)\exp(-i\omega t)$. Applying perturbation theory to the time dependent Schrödinger equation

$$\{T(r) + U(r) + \delta U(r)\exp(-i\omega t) - i\hbar\partial/\partial t\}\,\psi(r, t) = 0 \tag{7.23}$$

we assume the unperturbed orbital to be $|i\rangle$ and put

$$\psi(r, t) = |i\rangle \exp[-iE_i t/\hbar] + \sum_{k \neq i} a_{ik}|k\rangle \exp[-i(E_i + \hbar\omega)t/\hbar]. \tag{7.24}$$

Substituting Eq. (7.24) into Eq. (7.23) we obtain

$$\delta U(r)|i\rangle + \sum_{k \neq i} a_{ik}(E_k - E_i - \hbar\omega)|k\rangle = 0. \tag{7.25}$$

Multiplication with $|k\rangle$ from the left and integration yields the coefficients

$$a_{ik} = \langle k|\delta U(r)|i\rangle/(E_i + \hbar\omega - E_k). \tag{7.26}$$

The reaction potential caused by the perturbed one-electron orbital (7.24) is obtained by integrating the charge density together with the electronic interaction potential $V_{el}(r, s)$, yielding

$$V_{rct}(s) = \langle\psi|V_{el}(r, s)|\psi\rangle \tag{7.27}$$

and

$$V_{\text{rct}}(s) = \sum_{k \neq i} \left\{ \frac{\langle i| V_{\text{el}}(r, s)|k \rangle \langle k|\delta U(r)|i \rangle}{E_i + \hbar\omega - E_k} \exp(-i\omega t) \right.$$

$$\left. + \frac{\langle k| V_{\text{el}}(r, s)|i \rangle \langle i|\delta U^*(r)|k \rangle}{E_i + \hbar\omega - E_k} \exp(i\omega t) \right\}. \tag{7.28}$$

By requiring the perturbing potential to be real, i.e., by substituting

$$\delta U(r) \exp(-i\omega t) + \delta U^*(r) \exp(i\omega t) \tag{7.29}$$

and summing over all occupied electron orbitals $\psi(r, t)$, we find the total reaction potential of particle 1 at position s to be

$$V_{\text{rct}}(s) = \sum_{i,k} f_i(1 - f_k) \left\{ \frac{\langle i| V_{\text{el}}|k \rangle \langle k|\delta U|i \rangle}{E_i + \hbar\omega - E_k} \right.$$

$$\left. + \frac{\langle k| V_{\text{el}}|i \rangle \langle i|\delta U|k \rangle}{E_i - \hbar\omega - E_k} \right\} \exp(-i\omega t) + \text{c.c.} \tag{7.30}$$

Interchanging subscripts i and k in the terms containing $E_i - \hbar\omega - E_k$ yields

$$V_{\text{rct}}(s) = \sum_{i,k} \frac{f_i - f_k}{E_i + \hbar\omega - E_k} \langle i| V_{\text{el}}(r, s)|k \rangle \langle k|\delta U(r)|i \rangle$$

$$\cdot \exp(-i\omega t) + \text{c.c.} \tag{7.31}$$

Considering now two particles 1 and 2, we ask ourselves whether resonant states exist. We assume the reaction potential $V_{\text{rct}}(s)$ of particle 1 to be the external perturbation for particle 2, and the reaction potential of particle 2 to be the external perturbation for particle 1, yielding

$$\delta U_2(s) = \sum_{i,k} \frac{f_i - f_k}{E_i + \hbar\omega - E_k} \langle i| V_{\text{el}}|k \rangle \langle k|\delta U_1|i \rangle \tag{7.32}$$

$$\delta U_1(s) = \sum_{j,l} \frac{f_j - f_l}{E_j + \hbar\omega - E_l} \langle j| V_{\text{el}}|l \rangle \langle l|\delta U_2|j \rangle. \tag{7.33}$$

Equations (7.32) and (7.33) represent a set of homogeneous integral equations for the resonant potentials $\delta U_2(s)$, $\delta U_1(s)$ and the respective eigenfrequencies Ω_n. If we expand $V_{\text{el}}(r_i, r_j)$ into a Taylor series in $r_i - r_1$ and $r_j - r_2$ up to terms of the second order, assume the mean intensity of oscillations of the resonant potentials to obey Bose statistics, and apply the state density integration method described in Section 3.4, we find

$$\Delta E_{\text{dis}} = - (\hbar/4\pi i) \oint\limits_{-i\infty}^{+i\infty} d\omega \coth(\hbar\omega/2kT)$$
$$\cdot \operatorname{tr} \{ \mathbf{X}_1(\omega) \cdot V_1 V_2 V_{\text{el}} \cdot \mathbf{X}_2(\omega) \cdot V_1 V_2 V_{\text{el}} \} \ . \tag{7.34}$$

We recover the dispersion energy (7.22) obtained by applying time-independent perturbation theory to the Schrödinger equation, and find an additional factor $\coth(\hbar\omega/2kT)$ accounting for the intensity of the interacting fields.

7.3. Multipole Susceptibilities

The preceding investigations are not restricted to dipole interactions. On keeping higher than second order terms in the expansion of $W(r_i, r_j)$ with respect to $r_i - r_1$ and $r_j - r_2$, we obtain the quadrupole and octupole contributions to the orientation, induction, and dispersion energy. If we wish to treat these contributions separately, we have to require that an electron transition from state $|i\rangle$ to state $|k\rangle$ interferes either with the dipole or with the quadrupole (octupole) potential. The potentials and transitions must be properly diagonalized.

In the following we consider in particular the multipole susceptibilities of spherical particles. The spherical symmetry entails that all orbitals split into radial functions times spherical harmonics and that the reaction potential of the particle to an external perturbation shows the same symmetry as the latter. A free atom is a priori spherical. The inert gas atoms may be assumed spherical even in their respective lattices.

With $U_1(r)$, $U_2(r)$ depending on the radial coordinate r only, we may represent the solutions of Schrödinger equations (7.1), (7.2) by

$$|i\rangle = ((l-\lambda)!/(l+\lambda)!)^{1/2} |i\rangle_r P_l^\lambda(\cos\vartheta) \exp(i\lambda\varphi) \tag{7.35}$$

where the radial eigenvector $|i\rangle_r$ satisfies

$$\{ -(\hbar^2/2m)(d^2/dr^2 - l(l+1)/r^2) + U(r) - E_i \} r|i\rangle_r = 0 \ . \tag{7.36}$$

It is conveniently normalized to give

$$\langle i| r^2 |i\rangle_r = (2l+1)/4\pi \ . \tag{7.37}$$

To obtain the dispersion energy between the spherical particles 1 and 2 in general order, we ask, in a similar manner to our semiclassical treatment in Chapter 4, for their reaction to the external multipole potential

$$\delta U(r) = r^m P_m^\mu(\cos\vartheta) \exp(i\mu\varphi) \ . \tag{7.38}$$

Substituting the orbitals (7.35) and the potential (7.38) into $V_{\text{rct}}(s)$ according to Eq. (7.31), we note that the set of eigenvalues E_i of Eq. (7.36) does not depend on the rotation wave number λ. We may perform the summation over the orientation of orbitals $|i\rangle$ and $|k\rangle$ independent of the energy denominators $E_i + \hbar\omega - E_k$ and independent of the occupation numbers f_i, f_k. This summation is achieved by using the orthogonality relation

$$\sum_{\nu,\lambda} \frac{(n-\nu)!}{(n+\nu)!} \frac{(l-\lambda)!}{(l+\lambda)!} P_n^\nu e^{i\nu\varphi} P_l^\lambda e^{-i\lambda\varphi} \{P_n^\nu e^{-i\nu\varphi} P_m^\mu e^{i\mu\varphi} P_l^\lambda e^{i\lambda\varphi}\}$$
$$= (-1)^\mu e^{i\mu\varphi} \sum_\lambda P_n^{\lambda+\mu} P_l^{-\lambda} \{P_n^{-\lambda-\mu} P_m^\mu P_l^\lambda\} \tag{7.39}$$

$$= P_m^\mu e^{i\mu\varphi} \{P_n P_m P_l\} . \tag{7.40}$$

The braces in Eq. (7.39), (7.40) denote the integration over the solid angle $d\Omega = d\cos\vartheta\, d\varphi$. Equation (7.39) results from the orthogonality condition $\nu = \lambda + \mu$ with respect to the φ-integration. Equation (7.40) is conveniently proved by expanding $P_n^{-\lambda-\mu} P_l^\lambda$ within the Ω-integral in terms of Legendre functions $P_{n+l-2k}^{-\mu}$ of order $-\mu$, $k = 0, 1, \ldots, l$. The summation over λ for fixed k then brings about the respective term P_{n+l-2k}^μ before the Ω-integral. On replacing P_{n+l-2k}^μ by P_m^μ by means of the orthogonality relation of associated Legendre functions and summing over k according to

$$P_n P_l = \sum_{k=0}^l \left(\frac{(n+l-k)!}{(n-k)!(l-k)!k!}\right)^2 \frac{(2n-2k)!(2l-2k)!(2k)!}{(2n+2l-2k+1)!}$$
$$\cdot (2n+2l-4k+1) P_{n+l-2k} \tag{7.41}$$

we finally have Eq. (7.40).

By substituting Eq. (7.40) into Eq. (7.31), we find

$$V_{\text{rct}}(s) = \exp(-i\omega t) \sum_{(i,k)} \frac{f_i - f_k}{E_i + \hbar\omega - E_k} \langle i|r^2\{V_{\text{el}}(r,s) P_m^\mu e^{i\mu\varphi}\}|k\rangle_r$$
$$\cdot \langle k|r^{m+2}|i\rangle_r \{P_l P_m P_n\} + \text{c.c.} \tag{7.42}$$

with the summations over i, k now covering the degree l, n of the respective Legendre functions and the radial eigenvectors $|i\rangle_r$, $|k\rangle_r$.

Turning to the electron interaction potential $V_{\text{el}}(r, s)$, we note that it depends on the radial coordinates r, s and on the enclosed angle ϑ_{rs} only, due to the spherical symmetry of the particles under investigation. In particular, when s lies outside the region of non-zero charge density $\langle\psi\|\psi\rangle$, it equals the Coulomb potential $|r-s|^{-1}$. Using the general expansion

$$V_{\text{el}}(r, s) = \sum_{n=0}^\infty u_n(r, s) P_n(\cos\vartheta_{rs}) \tag{7.43}$$

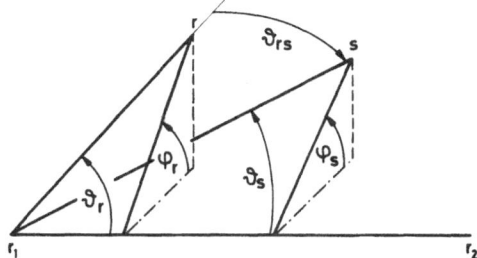

Fig. 28. Standard coordinates

with

$$V_{el}(r, s) = e^2/|r - s| = e^2 \sum_{n=0}^{\infty} (r^n/s^{n+1}) P_n(\cos \vartheta_{rs}) \tag{7.44}$$

in the exterior, and representing ϑ_{rs} in terms of the standard coordinates ϑ_r, φ_r and ϑ_s, φ_s according to Fig. 28 and addition theorem (4.12),

$$P_n(\cos \vartheta_{rs}) = \sum_{\nu=-n}^{+n} (-1)^{\nu} P_n^{-\nu}(\cos \vartheta_r) P_n^{\nu}(\cos \vartheta_s) e^{i\nu(\varphi_s - \varphi_r)} \tag{7.45}$$

we obtain

$$V_{rct}(s) = \exp(-i\omega t) 4\pi(2m+1)^{-1} P_m^{\mu}(\cos \vartheta_s) e^{i\mu\varphi_s} \sum_{(i,k)} \frac{f_i - f_k}{E_i + \hbar\omega - E_k}$$
$$\cdot \langle i|r^2 u_m(r, s)|k\rangle_r \langle k|r^{m+2}|i\rangle_r \{P_l P_m P_n\} + \text{c.c.} \tag{7.46}$$

The reaction potential $V_{rct}(s)$ exhibits the same symmetry behavior as the perturbing multipole potential $\delta U(r)$. In the exterior of the particle under consideration, where the Coulomb potential (7.44) may be substituted, we obtain

$$V_{rct}(s) = -\exp(-i\omega t) \Delta(m, j) s^{-(m+1)} P_m^{\mu}(\cos \vartheta_s) e^{i\mu\varphi_s} + \text{c.c.} \tag{7.47}$$

where

$$\Delta(m, j) = 4\pi e^2(2m+1)^{-1} \sum_{(i,k)} \frac{f_i - f_k}{E_i + \hbar\omega - E_k} |\langle i|r^{m+2}|k\rangle_r|^2 \{P_l P_m P_n\}. \tag{7.48}$$

We are able to define a multipole susceptibility $\Delta(m, j)$ of particle j in a similar manner to the semiclassical treatment presented in Chapter 4. On substituting the quantum mechanical susceptibilities (7.48) for the respective classical terms, we can find the dispersion energy between particles 1 and 2 on strictly similar lines to the semiclassical procedure.

7.4. Retardation

Having split the correlation of electron orbitals into the coupling with free multipole potentials and into the interaction via these multipole potentials, we may readily investigate retardation. All that has to be done is to replace the electrostatic multipole potentials by electro-dynamic four potentials. We showed in Section 5.2 that it is in accord with Lorentz gauge to equate to zero the electric potential. The retarded interaction via electric and magnetic modes is fully covered by the magnetic vector potential $A(r)$. An oscillating external potential $A_{ext}(r)$ causes a current-charge oscillation within the particle considered. The current-charge oscillation causes a reacting vector potential $A_{ret}(r)$ which in turn acts as external potential upon all surrounding particles.

Linearizing the Schrödinger equation in the presence of a perturbing vector potential $\delta A(r) \exp(-i\omega t)$ we have

$$\{T(r) + (ie\hbar/2mc)(\nabla \cdot \delta A + \delta A \cdot \nabla) \exp(-i\omega t)$$
$$+ U(r) - i\hbar\partial/\partial t\} \, \psi(r, t) = 0 \, . \tag{7.49}$$

Solving Eq. (7.49) by time dependent perturbation theory in a similar manner to that used in the presence of a perturbing electric potential, we find that the coefficients a_{ik} of the perturbed orbital $\psi(r, t)$ according to Eq. (7.24) are

$$a_{ik} = \frac{ie\hbar}{2mc} \frac{\langle k|\nabla \cdot \delta A + \delta A \cdot \nabla|i\rangle}{E_i + \hbar\omega - E_k} \, . \tag{7.50}$$

By assuming the total perturbing potential $A_{ext}(r, t)$ to be real, i.e. by substituting

$$A_{ext}(r, t) = \delta A \exp(-i\omega t) + \delta A^* \exp(i\omega t) \tag{7.51}$$

we find the change in particle density caused by the one-electron orbital $\psi(r, t)$ to equal

$$\delta(\psi^*\psi) = (ie\hbar/2mc) \exp(-i\omega t) \sum_{k \neq i} \left[\frac{\langle k|\nabla \cdot \delta A + \delta A \cdot \nabla|i\rangle}{E_i + \hbar\omega - E_k} |k\rangle\langle i| \right.$$
$$\left. + \frac{\langle i|\nabla \cdot \delta A + \delta A \cdot \nabla|k\rangle}{E_i - \hbar\omega - E_k} |i\rangle\langle k| \right] + \text{c.c.} \tag{7.52}$$

The particle current density in the presence of the external vector potential A_{ext} is given by

$$j = (2m)^{-1} \{\psi(i\hbar\nabla - (e/c) A_{ext}) \psi^* - \psi^*(i\hbar\nabla + (e/c) A_{ext}) \psi\} \, . \tag{7.53}$$

For the change in particle current density caused by the orbital $\psi(r, t)$ we find

$$\delta j = (i\hbar/2m)(ie\hbar/2mc)\exp(-i\omega t)\sum_{k\neq i}\left[\frac{\langle k|\boldsymbol{V}\cdot\delta\boldsymbol{A}+\delta\boldsymbol{A}\cdot\boldsymbol{V}|i\rangle}{E_i+\hbar\omega-E_k}\right.$$

$$\cdot(|k\rangle\boldsymbol{V}\langle i|-\boldsymbol{V}|k\rangle\langle i|)$$

$$\left.+\frac{\langle i|\boldsymbol{V}\cdot\delta\boldsymbol{A}+\delta\boldsymbol{A}\cdot\boldsymbol{V}|k\rangle}{E_i-\hbar\omega-E_k}(|i\rangle\boldsymbol{V}\langle k|-\boldsymbol{V}|i\rangle\langle k|)\right] \tag{7.54}$$

$$-(e/mc)\delta\boldsymbol{A}\exp(-i\omega t)|i\rangle\langle i|+\text{c.c.}$$

By summing Eqs. (7.52) and (7.54) over all occupied one-electron orbitals $\psi(r, t)$ located at the particle under investigation, we obtain the induced current-charge density

$$\boldsymbol{J}(r,t)/c=\exp(-i\omega t)(ie\hbar/2mc)^2\sum_{i,k}(f_i-f_k)\frac{\langle k|\boldsymbol{V}\cdot\delta\boldsymbol{A}+\delta\boldsymbol{A}\cdot\boldsymbol{V}|i\rangle}{E_i+\hbar\omega-E_k}$$

$$\cdot(|k\rangle\boldsymbol{V}\langle i|-\boldsymbol{V}|k\rangle\langle i|,(2m/i\hbar)|k\rangle\langle i|) \tag{7.55}$$

$$-\exp(-i\omega t)(e^2/mc^2)\sum_i f_i|i\rangle\langle i|(\delta\boldsymbol{A},0)+\text{c.c.}$$

The second term in $\boldsymbol{J}(r, t)$ guarantees that the continuity equation is satisfied, i.e. we have

$$(\boldsymbol{V},\partial/\partial t)\boldsymbol{J}(r,t)=0. \tag{7.56}$$

In order to find the reaction potential of the particle under investigation, we have to solve the Helmholtz wave equation in the presence of the current-charge density given by Eq. (7.55). Within the linear approximation used, we find $\boldsymbol{J}(r, t)$ to exhibit the same frequency as the external potential, yielding

$$(\boldsymbol{V}\cdot\boldsymbol{V}+K^2)(A/\mu,\varepsilon V/c)_{\text{ret}}=-4\pi\boldsymbol{J}(r,t)/c. \tag{7.57}$$

The Green's function corresponding to the inhomogeneous Helmholtz equation is the spherical Bessel function $y_0(K|r-s|)$ of the second kind, so that the integral form of Eq. (7.57) reads

$$(A/\mu,\varepsilon V/c)_{\text{ret}}=-\int d\boldsymbol{r}\,K\,y_0(K|r-s|)\boldsymbol{J}(r,t)/c. \tag{7.58}$$

From Eq. (7.58) we obtain both a magnetic reaction potential $A_{\text{ret}}(r, t)$ and an electric reaction potential $V_{\text{ret}}(r, t)$. They automatically satisfy the Lorentz gauge Eq. (5.5), i.e. by applying the $(\boldsymbol{V},\partial/\partial t)$ operator to Eq. (7.58), using $\boldsymbol{V}_s y_0(K|r-s|)=-\boldsymbol{V}_r\,y_0(K|r-s|)$ and integrating by parts we obtain

$$\boldsymbol{V}\cdot A/\mu+(\partial/\partial t)\varepsilon V/c=-\int d\boldsymbol{r}\,K\,y_0(K|r-s|)(\boldsymbol{V},\partial/\partial t)\boldsymbol{J}(r,t)/c=0. \tag{7.59}$$

One may conclude from Eq. (7.58) that, in contrast to our semiclassical findings in Chapter 5 and 6, it might not be permissible merely to investigate interactions via the magnetic vector potential. There is an electric potential as well. However, we can generally prove that when $V(r, t)$ satisfies the homogeneous Helmholtz equation, we may mutually cancel the electric and magnetic potentials

$$V(r, t) \quad \text{and} \quad A(r, t) = (i\omega/c)^{-1} \nabla V(r, t) \tag{7.60}$$

without causing an additional electric or magnetic field or an additional current-charge density and without violating the Lorentz gauge. In the exterior of the particles under investigation, i.e. in the region of vanishing charge density according to Eqs. (7.52), (7.55), it is possible to remove the electric potential $V(r, t)$ by changing the gauge.

7.5. Spherical Particles

Let us now turn to spherical particles and let us ask for their reaction to the external magnetic or electric multipole potential

$$\delta A(r) = (K^{-1} \nabla \times)^s r j_m(Kr) P_m^\mu(\cos \vartheta) \exp(i\mu\varphi). \tag{7.61}$$

Considering first the magnetic modes $s = 1$, it is convenient to shift the curl operator appearing in the vector potential $\delta A(r)$ to the orbitals $|i\rangle$, $|k\rangle$ by integrating by parts, yielding

$$\frac{1}{2}\langle k| \nabla \cdot \delta A + \delta A \cdot \nabla |i\rangle$$
$$= K^{-1} \int dr (\nabla \langle k| \times \nabla |i\rangle) \cdot r j_m(Kr) P_m^\mu(\cos \vartheta) \exp(i\mu\varphi). \tag{7.62}$$

By dissecting the orbitals $|i\rangle$, $|k\rangle$ into a radial function times a spherical harmonic according to Eq. (7.35) we obtain

$$\frac{1}{2}\langle k| \nabla \cdot \delta A + \delta A \cdot \nabla |i\rangle = K^{-1} \langle k|r j_m|i\rangle_r \left(\frac{(n-\nu)!}{(n+\nu)!} \frac{(l-\lambda)!}{(l+\lambda)!}\right)^{1/2}$$
$$\cdot \{P_m^\mu e^{i\mu\varphi} [(\partial/\partial\vartheta) P_n^\nu e^{-i\nu\varphi} (\sin\vartheta)^{-1} (\partial/\partial\varphi) P_l^\lambda e^{i\lambda\varphi}$$
$$- (\sin\vartheta)^{-1} (\partial/\partial\varphi) P_n^\nu e^{-i\nu\varphi} (\partial/\partial\vartheta) P_l^\lambda e^{i\lambda\varphi}]\}. \tag{7.63}$$

Substituting Eq. (7.63) into $J(r, t)$ according to Eq. (7.55), we may again sum over the orientation of orbitals $|i\rangle$, $|k\rangle$ independent of the radial behavior and independent of energy. Just as in the case of the scalar orthogonality relation (7.39), (7.40), we may prove the vector orthogonality relation

$$\sum_{\nu,\lambda} \frac{(n-\nu)!}{(n+\nu)!} \frac{(l-\lambda)!}{(l+\lambda)!} P_n^\nu e^{i\nu\varphi} \nabla P_l^\lambda e^{i\lambda\varphi}$$
$$\cdot \{P_m^\mu e^{i\mu\varphi} [(\partial/\partial\vartheta) P_n^\nu e^{-i\nu\varphi} (\sin\vartheta)^{-1} (\partial/\partial\varphi) P_l^\lambda e^{i\lambda\varphi}$$
$$- (\sin\vartheta)^{-1} (\partial/\partial\varphi) P_n^\nu e^{-i\nu\varphi} (\partial/\partial\vartheta) P_l^\lambda e^{i\lambda\varphi}]\}$$
$$= r^{-1} (0, (\sin\vartheta)^{-1} \partial/\partial\varphi, -\partial/\partial\vartheta) P_m^\mu e^{i\mu\varphi} \{P_l^1 P_m^{-1} P_n^1/\sin\vartheta\}. \tag{7.64}$$

Hence,

$$J(r, t)/c = - \exp(- i\omega t)(K^{-1} \boldsymbol{V} \times \boldsymbol{r} P_m^\mu(\cos \vartheta) e^{i\mu\varphi}, 0)\langle i|_r$$

$$\cdot \left[(e\hbar/2mc)^2 \sum_{(i,k)} \frac{f_i - f_k}{E_i + \hbar\omega - E_k} \langle k|rj_m|i\rangle_r r^{-1} |k\rangle_r \{P_l^1 P_m^{-1} P_n^1/\sin \vartheta\} \right.$$

$$\left. + (e^2/mc^2) \sum_{(i)} f_i j_m |i\rangle_r \right] + \text{c.c.} \tag{7.65}$$

The induced current density is parallel to the external vector potential $\delta A(r)$, whereas no oscillations of the charge density arise.

In order to obtain the reaction potential caused by the current density (7.65), we insert Eq. (7.65) into expression (7.58). It is then convenient to shift the $\boldsymbol{V} \times \boldsymbol{r}$ operator from $J(r, t)$ to $y_0(K|r - s|)$ by integrating by parts, and to replace $\boldsymbol{V}_r \times \boldsymbol{r} y_0(K|r - s|)$ by $- \boldsymbol{V}_s \times \boldsymbol{s} y_0(K|r - s|)$ using the fact that $y_0(K|r - s|)$ is a function of the separation $|r - s|$. Finally, applying the addition theorem for spherical Bessel functions of order zero, Eq. (10.1.46) in Ref. [1], which for the situation shown in Fig. 28 reads

$$y_0(K|r - s|) = \sum_{n=0}^{\infty} (2n + 1) y_n(Ks) j_n(Kr) P_n(\cos \vartheta_{rs}) \tag{7.66}$$

$(s > r)$, and using Eq. (7.45) we obtain

$$A_{\text{rct}}(r, t) = - \exp(- i\omega t) \Delta_{11}(m, j)(K^{-1} \boldsymbol{V} \times) s\, y_m(Ks)$$

$$\cdot P_m^\mu(\cos \vartheta_s) e^{i\mu\varphi_s} + \text{c.c.} \tag{7.67}$$

where

$$\Delta_{11}(m, j)/4\pi\mu K = (e\hbar/mc)^2 \sum_{(i,k)} \frac{f_i - f_k}{E_i + \hbar\omega - E_k} |\langle k|rj_m|i\rangle_r|^2$$

$$\cdot \{P_l^1 P_m^{-1} P_n^1/\sin \vartheta\} \tag{7.68}$$

$$- (e^2/mc^2) \sum_{(i)} f_i \langle i|(rj_m)^2 |i\rangle_r .$$

The reaction potential of the spherical particle under investigation caused by the ingoing magnetic mode (7.61) is the potential (7.67) of the corresponding outgoing magnetic mode. Within the linear approximation used, we can define a multipole susceptibility $\Delta_{11}(m, j)$ between the amplitudes of the ingoing and the outgoing modes in the same way as in our semiclassical treatment in Chapters 5 and 6.

Let us now inquire about the effect of the external electric modes with a vector potential given by Eq. (7.61) and $s = 2$. By separating the orbitals $|i\rangle, |k\rangle$ into a radial function times a spherical harmonic according to Eq. (7.35), we obtain in a similar manner to the case of the magnetic modes

$$\langle k| \boldsymbol{V} \cdot \delta \boldsymbol{A} + \delta \boldsymbol{A} \cdot \boldsymbol{V} |i\rangle = \left(\frac{(n-v)!}{(n+v)!} \frac{(l-\lambda)!}{(l+\lambda)!} \right)^{1/2} \{P_m^\mu e^{i\mu\varphi} P_n^v e^{-iv\varphi} P_l^\lambda e^{i\lambda\varphi}\} \quad (7.69)$$

$$\cdot K^{-2} \langle k|m(m+1)[(\partial/\partial r)rj_m + rj_m(\partial/\partial r)] + [l(l+1) - n(n+1)](\partial rj_m/\partial r)|i\rangle_r .$$

Substituting Eq. (7.69) into $J(r, t)$ according to Eq. (7.55) and summing over the orientation of the orbitals $|i\rangle, |k\rangle$ regardless of the radial behavior and of energy, we may apply orthogonality relation (7.40) to the summation of the radial component of the current density and to that of the charge density. An analogous summation of the angular components of the current density is possible on the basis of

$$\sum_{v,\lambda} \frac{(n-v)!}{(n+v)!} \frac{(l-\lambda)!}{(l+\lambda)!} P_n^v e^{iv\varphi} (\partial/\partial\vartheta, (\sin\vartheta)^{-1}\partial/\partial\varphi)$$

$$\cdot P_l^\lambda e^{-i\lambda\varphi} \{P_m^\mu e^{i\mu\varphi} P_n^v e^{-iv\varphi} P_l^\lambda e^{i\lambda\varphi}\} \quad (7.70)$$

$$= (\partial/\partial\vartheta, (\sin\vartheta)^{-1}\partial/\partial\varphi) P_m^\mu e^{i\mu\varphi} \{-P_l^1 P_m^{-1} P_n\} .$$

The integrals $\{-P_l^1 P_m^{-1} P_n\}$, $\{-P_l P_m^{-1} P_n^1\}$, and $\{P_l P_m P_n\}$ over the solid angle $d\Omega = d\cos\vartheta\,d\varphi$ are related by

$$\{-P_l^1 P_m^{-1} P_n\} + \{-P_l P_m^{-1} P_n^1\} = \{P_l P_m P_n\} \quad (7.71)$$

$$\{-P_l^1 P_m^{-1} P_n\} - \{-P_l P_m^{-1} P_n^1\} = \frac{l(l+1) - n(n+1)}{m(m+1)} \{P_l P_m P_n\} \quad (7.72)$$

as is proved in Section 7.6. Hence,

$$J(r, t)/c = -\exp(-i\omega t)(e\hbar/2mc)^2 K^{-2} \sum_{(i,k)} \frac{f_i - f_k}{E_i + \hbar\omega - E_k}$$

$$\cdot \langle k|m(m+1)[(\partial/\partial r)rj_m + rj_m\partial/\partial r] + [l(l+1) - n(n+1)](\partial rj_m/\partial r)|i\rangle_r$$

$$\cdot (|k\rangle_r (\partial/\partial r)\langle i|_r - (\partial/\partial r)|k\rangle_r \langle i|_r, |k\rangle_r \langle i|_r \frac{l(l+1) - n(n+1)}{m(m+1)} r^{-1}\partial/\partial\vartheta, \quad (7.73)$$

$$|k\rangle_r \langle i|_r \frac{l(l+1) - n(n+1)}{m(m+1)} (r\sin\vartheta)^{-1}\partial/\partial\varphi, (2m/i\hbar)|k\rangle_r \langle i|_r) P_m^\mu e^{i\mu\varphi}$$

$$- \exp(-i\omega t)(e^2/mc^2) K^{-2} \sum_{(i)} f_i |i\rangle_r \langle i|_r$$

$$\cdot [r^{-2}m(m+1), r^{-1}\partial^2/\partial r\partial\vartheta, (r\sin\vartheta)^{-1}\partial^2/\partial r\partial\varphi, 0] rj_m(Kr) P_m^\mu e^{i\mu\varphi} .$$

In contrast to our findings in the presence of an external magnetic mode we obtain a non-vanishing charge density.

We find the reaction potential caused by the current-charge density (7.73) by substituting $J(r, t)/c$ into Eq. (7.58). Again representing $y_0(K|r - s|)$ by spherical Bessel functions times spherical harmonics according to Eqs. (7.66) and (7.45), we may perform the integration over the solid angle $d\Omega$ by using

$$(4\pi)^{-1} \{y_0(K|r - s|) (a(r), b(r)\partial/\partial\vartheta, b(r)(\sin\vartheta)^{-1}\partial/\partial\varphi) P_m^\mu(\cos\vartheta) e^{i\mu\varphi}\}$$
$$= (K^{-1} V \times)^2 s y_m(Ks) P_m^\mu(\cos\vartheta_s) e^{i\mu\varphi_s}[a(r)r^{-1} + b(r) r^{-1}(d/dr)r]j_m(Kr)$$
$$+ K^{-2} V y_m(Ks) P_m^\mu(\cos\vartheta_s) e^{i\mu\varphi_s}[a(r)d/dr + b(r)r^{-1}m(m+1)]j_m(Kr). \quad (7.74)$$

According to Eq. (7.74) we find the vector potential $A_{rct}(r, t)$ to contain terms $(K^{-1} V \times)^2 s y_m(Ks) P_m(\cos\vartheta_s) \exp(i\mu\varphi_s)$ and $V y_m(Ks) P_m(\cos\vartheta_s)$ $\cdot \exp(i\mu\varphi_s)$ as well. The first term represents an outgoing electric mode, whereas the latter term can be cancelled with the resulting electric potential

$$(\varepsilon/4\pi) V_{rct}(r, t) = -\exp(-i\omega t) (e^2\hbar/2mc)K^{-1} y_m(Ks) P_m^\mu(\cos\vartheta_s) e^{i\mu\varphi_s}$$

$$\cdot \sum_{(i,k)} \frac{f_i - f_k}{E_i + \hbar\omega - E_k} \langle k|m(m+1) [(d/dr)rj_m + rj_m d/dr] \quad (7.75)$$

$$+ [l(l+1) - n(n+1)] (drj_m/dr)|i\rangle_r \langle i|r^2 j_m|k\rangle_r + \text{c.c.}$$

by changing the gauge according to Eq. (7.60). Hence,

$$A_{rct}(r,t) = -\exp(-i\omega t)\Delta_{22}(m,j)(K^{-1} V \times)^2 s y_m(Ks) P_m^\mu(\cos\vartheta_s) e^{i\mu\varphi_s} + \text{c.c.}$$
$$(7.76)$$

where

$$(4\pi\mu)^{-1} K \Delta_{22}(m, j) = (e\hbar/2mc)^2 \sum_{(i,k)} \frac{f_i - f_k}{E_i + \hbar\omega - E_k} \frac{\{P_l P_m P_n\}}{m(m+1)}$$

$$\cdot |\langle k|m(m+1)[(d/dr)rj_m + rj_m d/dr] + [l(l+1) - n(n+1)](drj_m/dr)|i\rangle_r|^2 \quad (7.77)$$

$$- (e^2/mc^2) \sum f_i \langle i|m(m+1) j_m^2 + (drj_m/dr)^2 |i\rangle_r.$$

The reaction potential of the spherical particle under investigation due to the ingoing electric mode (7.61) is the potential (7.76) of the corresponding outgoing mode. As in the case of an external magnetic mode, we obtain the respective electric multipole susceptibility $\Delta_{22}(m, j)$ in terms of the radial electron orbitals $|i\rangle_r, |k\rangle_r$.

By taking the nonretarded limit of the electric multipole susceptibility $\Delta_{22}(m, j)$, we recover the electrostatic multipole susceptibility (7.48) except for the factor $K^{2m+1} 2^{2m}(m-1)!(m+1)!/(2m)!(2m+1)!$, which compensates for the differing definitions of the ingoing and outgoing potentials given by Eqs. (7.38), (7.47), and (7.61), (7.76) in the two limits.

By substituting the quantum theoretical susceptibilities $\Delta_{11}(m,j)$, $\Delta_{22}(m,j)$ according to Eqs. (7.68), (7.77) into the calculations presented in Section 6, we find the retarded dispersion energy between two particles 1 and 2 in spite of having used the nonretarded Schrödinger equation. Owing to the separate treatment of the coupling of particles 1 and 2 to the electrodynamic field and of their interaction via this field the propagation time is exclusively left to the latter. The coupling of particles 1 and 2 to the electrodynamic field is completely covered by the coupling to the transverse harmonic modes introduced in Section 5.2, and thus is not affected by retardation at all.

The main difference between the quantum theoretical susceptibilities $\Delta_{ss}(m,j)$ according to Eqs. (7.68), (7.77) and the respective classical quantities (5.24), (5.25) is the dependence of the former on temperature via the occupation numbers f_i. f_k. The retarded dispersion energy between the particles considered, in accordance with the underlying electron-photon-exchange interaction, depends both on Bose statistics and on Fermi statistics.

7.6. Integrals of Legendre Functions

In the preceding sections we introduced several integrals and orthogonality relations of associated Legendre functions, which we shall prove in the following:

To find the integrals $\{P_l P_m P_n\}$, $\{P_l^1 P_m^{-1} P_n^1/\sin\vartheta\}$ necessary in the orthogonality relations (7.40), (7.64), we consider the general expansion

$$P_n^v(\cos\vartheta)\, P_l^v(\cos\vartheta)/(-\sin\vartheta)^v$$

$$= \frac{v!}{2^v(2v)!} \sum_{k=0}^{l-v} \frac{(2n-2k)!}{(n-k)!(n-v-k)!}\; \frac{(2l-2k)!}{(l-k)!(l-v-k)!}\; \frac{(2k+2v)!}{(k+v)!k!} \quad (7.78)$$

$$\cdot \frac{(n+l-k)!(n+l-v-k)!}{(2n+2l-2v-2k+1)!}\; \frac{(n+l-2v-2k)!}{(n+l-2k)!}$$

$$\cdot (2n+2l-2v-4k+1)\, P_{n+l-v-2k}^v(\cos\vartheta)\,.$$

Equation (7.78) is conveniently verified by using the initial value $l=v$

$$P_n^v(\cos\vartheta)\, P_v^v(\cos\vartheta)/(-\sin\vartheta)^v = ((2v)!/2^v\, v!)\, P_n^v(\cos\vartheta) \quad (7.79)$$

and complete induction with respect to l by means of

$$\cos\vartheta \cdot P_n^v P_l^v = P_n^v\left(\frac{l-v+1}{2l+1}\, P_{l+1}^v + \frac{l+v}{2l+1}\, P_{l-1}^v\right)$$

$$+ P_l^v\left(\frac{n-v+1}{2n+1}\, P_{n+1}^v + \frac{n+v}{2n+1}\, P_{n-1}^v\right). \quad (7.80)$$

Multiplication of Eq. (7.78) with $P_m^{-\nu}(\cos\vartheta)$ and integration over the solid angle $d\Omega = d\cos\vartheta\, d\varphi$ yields

$$(4\pi)^{-1}\{P_l^\nu P_m^{-\nu} P_n^\nu/(\sin\vartheta)^\nu\}$$

$$= \frac{\nu!}{2^\nu(2\nu)!}\,\frac{(m-\nu)!}{(m+\nu)!}\,\frac{[(l+m+n+\nu)/2]!\,[(l+m+n-\nu)/2]!}{(l+m+n-\nu+1)!} \qquad (7.81)$$

$$\cdot\,\frac{(l+m+n+\nu)!}{[(l+m+n+\nu)/2]!\,[(l+m+n-\nu)/2]!}$$

$$\cdot\,\frac{(m+n-l+\nu)!}{[(l+m+n+\nu)/2]!\,[(l+m+n-\nu)/2]!}$$

$$\cdot\,\frac{(n+l-m+\nu)!}{[(l+m+n+\nu)/2]!\,[(l+m+n-\nu)/2]!}$$

In orthogonality relations (7.40), (7.64), we need the integral (7.81) in particular for $\nu = 0, 1$. Similarly, Eq. (7.41) is recovered from Eq. (7.78) by putting $\nu = 0$. Eqs. (7.71), (7.72) are conveniently proved by sucessively applying Eqs. (8.2.5) and (8.5.5) in Ref. [1] to give

$$\{-P_l^1 P_m^{-1} P_n\} = [m(m+1)]^{-1}\{P_l^1 P_m^1 P_n\}$$

$$= [m(m+1)]^{-1}\left\{P_l^1 P_m^1\left(\frac{(n+2)(n+1)}{2n+1}P_{n+1}^{-1} - \frac{n(n-1)}{2n+1}P_{n-1}^{-1}\right)/\sin\vartheta\right\}. \qquad (7.82)$$

We obtain two integrals of the type given in Eq. (7.81) with $\nu = 1$ and find

$$\{-P_l^1 P_m^{-1} P_n\} = \frac{m(m+1)+l(l+1)-n(n+1)}{2m(m+1)}\{P_l P_m P_n\}. \qquad (7.83)$$

In order to treat the integral (7.74) containing the Green's function $y_0(K|r-s|)$, we first turn from spherical to rectangular coordinates. Assuming the x-axis parallel to the direction $\vartheta = 0$ as shown in Fig. 29 we have

$$x = r\cos\vartheta\,; \qquad y = r\sin\vartheta\cos\varphi\,; \qquad z = r\sin\vartheta\sin\varphi \qquad (7.84)$$

and

$$(a(r), b(r)\,\partial/\partial\vartheta, b(r)(\sin\vartheta)^{-1}\partial/\partial\varphi)\,P_m^\mu(\cos\vartheta)\,e^{i\mu\varphi} \qquad (7.85)$$

$$= [a\cos\vartheta - b\sin\vartheta\partial/\partial\vartheta, e^{i\varphi}(a\sin\vartheta + b\cos\vartheta\partial/\partial\vartheta + ib(\sin\vartheta)^{-1}\partial/\partial\varphi)]$$

$$\cdot P_m^\mu(\cos\vartheta)\,e^{i\mu\varphi}\,.$$

Vectors in spherical and rectangular coordinates are distinguished by paranthesis and brackets, respectively. In rectangular coordinates we write down the x and the $y+iz$ components only. The $y-iz$ component is complex conjugate to the latter.

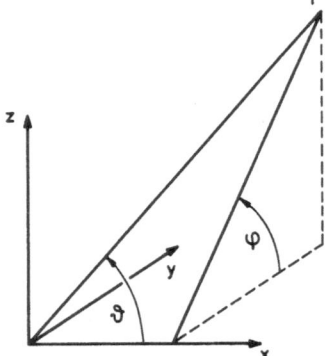

Fig. 29. Rectangular coordinates

Generally adapting the order of the Legendre polynomials in Eq. (7.85) to the rotation wave numbers μ, $\mu \pm 1$, we find according to Eqs. (8.5.1) to (8.5.5) in Ref. [1]

$$(a(r), b(r)\partial/\partial\vartheta, b(r)\,(\sin\vartheta)^{-1}\partial/\partial\varphi)\,P_m^\mu(\cos\vartheta)\,e^{i\mu\varphi}$$

$$= \frac{a(r) - m\,b(r)}{2m+1} [(m-\mu+1)\,P_{m+1}^\mu\,e^{i\mu\varphi},\,P_{m+1}^{\mu+1}\,e^{i(\mu+1)\varphi}] \tag{7.86}$$

$$+ \frac{a(r) + (m+1)\,b(r)}{2m+1} [(m+\mu)\,P_{m-1}^\mu\,e^{i\mu\varphi},\,-P_{m-1}^{\mu+1}\,e^{i(\mu+1)\varphi}].$$

By multiplying Eq. (7.86) with $y_0(K|r-s|)$, representing the latter in terms of spherical harmonics according to Eqs. (7.66) and (7.45), and integrating over the solid angle $d\Omega = d\cos\vartheta\,d\varphi$ we obtain

$$(4\pi)^{-1}\{y_0(K|r-s|)\,(a(r), b(r)\partial/\partial\vartheta, b(r)\,(\sin\vartheta)^{-1}\partial/\partial\varphi)\,P_m^\mu(\cos\vartheta)\,e^{i\mu\varphi}\}$$

$$= \frac{a(r) - m\,b(r)}{2m+1}\,j_{m+1}(Kr)\,y_{m+1}(Ks)\,[(m-\mu+1)$$

$$\cdot P_{m+1}^\mu\,e^{i\mu\varphi},\,P_{m+1}^{\mu+1}\,e^{i(\mu+1)\varphi}] \tag{7.87}$$

$$+ \frac{a(r) + (m+1)\,b(r)}{2m+1}\,j_{m-1}(Kr)\,y_{m-1}(Ks)\,[(m+\mu)$$

$$\cdot P_{m-1}^\mu\,e^{i\mu\varphi},\,-P_{m-1}^{\mu+1}\,e^{i(\mu+1)\varphi}]$$

with the polar coordinates ϑ, φ now referring to s rather than to r.

Now. consider the potentials

$$\nabla y_m(Ks)\, P_m^\mu(\cos\vartheta)\, e^{i\mu\varphi}$$

$$= (\partial/\partial s,\, s^{-1}\partial/\partial\vartheta,\, (s\sin\vartheta)^{-1}\partial/\partial\varphi)\, y_m(Ks)\, P_m^\mu(\cos\vartheta)\, e^{i\mu\varphi}$$

$$= -K(2m+1)^{-1} y_{m+1}(Ks)[(m-\mu+1)P_{m+1}^\mu e^{i\mu\varphi},\, P_{m+1}^{\mu+1} e^{i(\mu+1)\varphi}]$$

$$+ K(2m+1)^{-1} y_{m-1}(Ks)[(m+\mu)P_{m-1}^\mu e^{i\mu\varphi},\, -P_{m-1}^{\mu+1} e^{i(\mu+1)\varphi}] \tag{7.88}$$

and

$$(\nabla\times)^2 s\, y_m(Ks)\, P_m^\mu(\cos\vartheta)\, e^{i\mu\varphi}$$

$$= (s^{-1}m(m+1),\, s^{-1}(\partial^2/\partial s\partial\vartheta)s,\, (s\sin\vartheta)^{-1}(\partial^2/\partial s\partial\varphi)s)$$

$$\cdot y_m(Ks)\, P_m^\mu(\cos\vartheta)\, e^{i\mu\varphi} \tag{7.89}$$

$$= Km(2m+1)^{-1} y_{m+1}(Ks)[(m-\mu+1)P_{m+1}^\mu e^{i\mu\varphi},\, P_{m+1}^{\mu+1} e^{i(\mu+1)\varphi}]$$

$$+ K(m+1)(2m+1)^{-1} y_{m-1}(Ks)[(m+\mu)P_{m-1}^\mu e^{i\mu\varphi},\, -P_{m-1}^{\mu+1} e^{i(\mu+1)\varphi}] .$$

It is obvious from Eqs. (7.87)–(7.89) that a linear representation of the integral (7.87) in terms of the potentials (7.88), (7.89) exists. We obtain Eq. (7.74) by applying Eqs. (10.1.19) and (10.11.20) in Ref. [1].

8. Electrons and Photons

8.1. Second Quantization

We have now extended the macroscopic approach to van der Waals attraction to cover the quantization of matter. Rather than calculating the reaction potential of the particles under investigation on the basis of their macroscopic permeabilities, we applied Schrödinger's equation. The electron transitions caused by a perturbing ingoing electric or magnetic mode induce the respective outgoing electric or magnetic mode. We can define multipole susceptibilities $\Delta_{ss}(m, j)$ of particle j in a similar manner as in the macroscopic investigations. By representing the outgoing modes at particle j in terms of ingoing modes at the surrounding particles on the basis of the addition theorem (5.36), we obtain a linear secular system for the possible electromagnetic eigenvectors. The van der Waals free energy of the system under investigation results by providing all electromagnetic eigenvectors with the free energy $kT\ln 2\sin(\hbar\omega/2kT)$ of Bosons.

In contrast to earlier approaches, we do not use the electrostatic interaction potential between electrons located at different particles and a perturbation theory with respect to free electrodynamic modes. We rather equate the electric interaction potential to zero and attribute the interaction exclusively to the transverse electric and magnetic modes. In

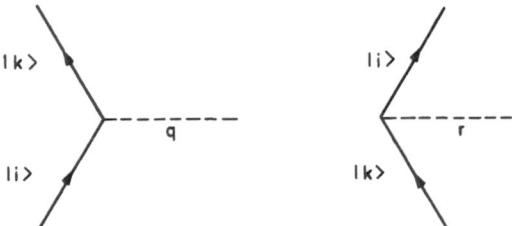

Fig. 30. Electron-photon interactions

other words, by diagonalizing the free eigenvectors of the Helmholtz equation with respect to the perturbing local electron transitions, we reduce both the mathematical and the physical effort. In spite of the fact that Schrödinger's equation is non-relativistic, it is permissible to apply it to investigations on the retarded dispersion energy. We use it for calculating the reaction potential of the single particles, but leave the propagation of fields to the Helmholtz equation.

Dealing with electron transitions which cause classical electrodynamic fields and classical fields which in turn cause electron transitions, we now ask about the effect of quantizing the electrodynamic field. The actual mechanism underlying van der Waals attraction is a photon exchange between the electrons localized at the particles under investigation. The electrodynamic field considered so far is a substitute for the average effect of the photons. The electron transition from orbital $|i\rangle$ to orbital $|k\rangle$ is coupled to the emission or absorption of an ingoing photon q, while the inverse transition from $|k\rangle$ to $|i\rangle$ causes the emission or absorption of the outgoing photon r, see Fig. 30.

The Hamiltonian describing this electron-photon interaction is conveniently given in terms of second quantization

$$H = \sum_i E_i c_i^+ c_i + \sum_q \tfrac{1}{2}\hbar\omega_q(a_q a_q^+ + a_q^+ a_q)$$
$$+ \sum_{ikq} \{U^+(kiq) a_q^+ + U^-(kiq) a_q\} c_k^+ c_i. \tag{8.1}$$

The first term in Eq. (8.1) represents the energy of the electrons. The operators c_i^+ and c_i are the creation and annihilation operators for Fermions. By applying the product $c_i^+ c_i$ to an arbitrary many electron orbital, we obtain the occupation number $f_i = 0, 1$ of state $|i\rangle$, the application of the commutated product $c_i c_i^+$ renders $1 - f_i$.

We have

$$c_i^+ c_i = f_i; \qquad c_i c_i^+ = 1 - f_i \tag{8.2}$$

and the anticommutation relation

$$c_i^+ c_i + c_i c_i^+ = 1 . \tag{8.3}$$

The second term in Eq. (8.1) represents the energy of the photons. The operators a_q^+ and a_q are the creation and annihilation operators for Bosons. They satisfy the commutation relation

$$[a_q, a_q^+] \equiv a_q a_q^+ - a_q^+ a_q = 1 \tag{8.4}$$

i.e. by applying the products $a_q^+ a_q$ and $a_q a_q^+$ to an arbitrary many photon orbital we obtain the occupation numbers

$$a_q^+ a_q = n_q ; \qquad a_q a_q^+ = n_q + 1 . \tag{8.5}$$

The third term in Eq. (8.1) gives the electron-photon interaction. An electron in orbital $|i\rangle$ is annihilated, an electron in orbital $|k\rangle$ is created. This electron transition is coupled to the emission or absorption of a photon q. The respective coupling parameters are denoted by $U^+(kiq)$ and $U^-(kiq)$. They are independent of the position of photon emission and absorption only if the photons correspond to planar modes.

In the present investigations we consider the photons to be ingoing electric and magnetic modes centered at an arbitrary position. They are absorbed or emitted by interaction with an electron transition, while the inverse transition gives rise to the corresponding outgoing electric or magnetic modes. The outgoing modes can not be considered to be photons, they satisfy the homogeneous Helmholtz equation only in the region of vanishing electron density. They rather are wave packets of photons, which can be expanded in terms of ingoing modes in the exterior. The products of the coupling parameters $U^{\pm}(kiq)\,U^{\mp}(ljq)$ describing emission and absorption of a photon q by electrons located at different particles 1 and 2 thus depend on the separation r_{21} in the same way as the transposition parameters $V_{mn}^{\mu}(Kr_{21})$, $W_{mn}^{\mu}(Kr_{21})$ appearing in the addition theorem (5.36).

8.2. Fourth Order Perturbation Theory

Van der Waals attraction is due to a photon exchange between electrons located at different particles 1 and 2, i.e. the lowest order contribution to the attraction arises from the graph shown in Fig. 31. In order to obtain the energy of attraction, we have to apply at least fourth order perturbation theory. Since in Sections 3.3 and 7.1 we have seen how to generalize the results of finite perturbation theory by complex integration techniques, we shall be able to extend the present findings in a similar manner.

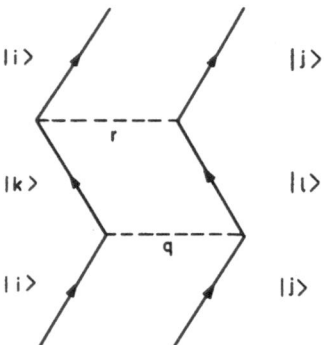

Fig. 31. Photon-exchange interaction

In the absence of the electron-photon interaction term in Eq. (8.1), we may represent the eigenvectors of

$$(H - E)\psi = 0 \tag{8.6}$$

by quoting the occupation numbers f_i and n_q of all electron states and all photon states

$$\psi_0 = |f_i f_j \ldots; n_q n_r \ldots\rangle. \tag{8.7}$$

For the energy of this unperturbed orbital we obtain

$$E_0 = \sum_i E_i f_i + \sum_q \hbar\omega_q (n_q + \tfrac{1}{2}). \tag{8.8}$$

The interaction term in the Hamiltonian (8.1) causes electron transitions and photon emission or absorption. We put

$$\psi = \psi_0 + \psi_1 + \psi_2 + \cdots \tag{8.9}$$

and

$$E = E_0 + E_2 + E_4 + \cdots. \tag{8.10}$$

The subscripts in Eqs. (8.9), (8.10) denote the order of the perturbation. Insertion into Eq. (8.6) yields

$$\psi_1 = \sum_{ikq} \left\{ \frac{U^+(kiq)}{E - E_0 - (E_k - E_i + \hbar\omega_q)} a_q^+ \right. $$
$$\left. + \frac{U^-(kiq)}{E - E_0 - (E_k - E_i - \hbar\omega_q)} a_q \right\} c_k^+ c_i \psi_0 \tag{8.11}$$

and

$$E_2 = \sum_{ikq} f_i(1 - f_k) \left\{ \frac{U^+(kiq)\, U^-(ikq)}{E - E_0 - (E_k - E_i + \hbar\omega_q)} (n_q + 1) \right.$$
$$\left. + \frac{U^-(kiq)\, U^+(ikq)}{E - E_0 - (E_k - E_i - \hbar\omega_q)} n_q \right\}. \tag{8.12}$$

The energy terms of order two result from emission and absorption of one photon by one electron. These terms represent part of the self-energy of the electrons and the photons. Since they do not depend on the separation of particles 1 and 2, we may cancel them by renormalization. Expanding the energy denominators in Eq. (8.12) with respect to $E - E_0$, we put

$$\langle E_i \rangle = E_i - \tfrac{1}{2} \sum_{kq} (1 - f_k) \left\{ \frac{U^+(kiq)U^-(ikq)}{E_k - E_i + \hbar\omega_q} - \frac{U^-(kiq)U^+(ikq)}{E_k - E_i - \hbar\omega_q} \right\} \tag{8.13}$$

$$\langle \hbar\omega_q \rangle = \hbar\omega_q - \sum_{ik} f_i(1 - f_k) \left\{ \frac{U^+(kiq)U^-(ikq)}{E_k - E_i + \hbar\omega_q} + \frac{U^-(kiq)U^+(ikq)}{E_k - E_i - \hbar\omega_q} \right\}. \tag{8.14}$$

Substituting the renormalized expressions (8.13) and (8.14) for the non-renormalized quantities in Eq. (8.8), we correctly recover the energy terms of order two.

Turning to the second order perturbation terms for the orbitals and to the fourth order perturbation terms for the energy we find

$$\psi_2 = {\sum_{jlr}}' \sum_{ikq} \left\{ \left(\frac{U^+(ljr)}{E - E_0 - (E_l - E_j + \hbar\omega_r + E_k - E_i + \hbar\omega_q)} a_r^+ \right. \right.$$
$$\left. + \frac{U^-(ljr)}{E - E_0 - (E_l - E_j - \hbar\omega_r + E_k - E_i + \hbar\omega_q)} a_r \right) \tag{8.15}$$
$$\left. \cdot \frac{U^+(kiq)}{E - E_0 + (E_k - E_i + \hbar\omega_q)} a_q^+ + \cdots \right\} c_l^+ c_j c_k^+ c_i \psi_0$$

and

$$E_4 = - \sum_{ik} f_i(1 - f_k) \sum_{jl} f_j(1 - f_l) \sum_{q\pm} (n_q + \tfrac{1}{2} \pm \tfrac{1}{2}) \sum_{r\pm} (n_r + \tfrac{1}{2} \pm \tfrac{1}{2})$$

$$\cdot \left\{ \left[\frac{U^\mp(jlr)U^\mp(ljq)}{E_l - E_j \pm \hbar\omega_r} + \frac{U^\mp(jlq)U^\mp(ljr)}{E_l - E_j \pm \hbar\omega_q} \right] \frac{U^\pm(ikr)U^\pm(kiq)}{(\pm\hbar\omega_q \pm \hbar\omega_r)(E_k - E_i \pm \hbar\omega_q)} \right.$$

$$+ \left[\frac{U^\mp(jlr)U^\pm(ikr)}{E_l - E_j \pm \hbar\omega_r} + \frac{U^\mp(ikr)U^\pm(jlr)}{E_k - E_i \pm \hbar\omega_r} \right] \frac{U^\mp(ljq)U^\pm(kiq)}{(E_k - E_i + E_l - E_j)(E_k - E_i \pm \hbar\omega_q)}$$

$$+ \left[\frac{U^\mp(jlq)U^\mp(ikr)}{E_l - E_j \pm \hbar\omega_q} + \frac{U^\mp(ikr)U^\mp(jlq)}{E_k - E_i \pm \hbar\omega_r} \right] \tag{8.16}$$

$$\left. \cdot \frac{U^\pm(ljr)U^\pm(kiq)}{(E_k - E_i \pm \hbar\omega_q + E_l - E_j \pm \hbar\omega_r)(E_k - E_i \pm \hbar\omega_q)} \right\}.$$

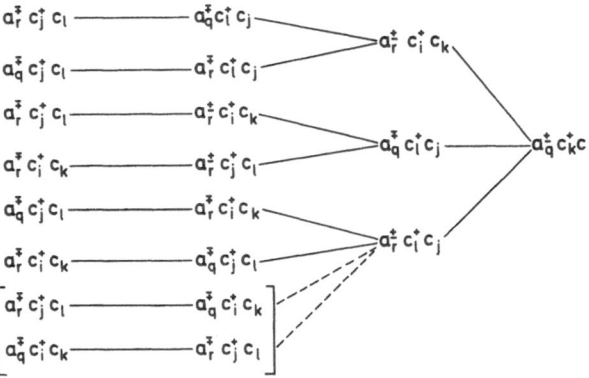

Fig. 32. Fourth order terms

The dots in Eq. (8.15) indicate the terms arising by absorption of a photon q. Similarly, the summations over $\pm$ in Eq. (8.16) extend to emission of photons q, r prior to absorption and absorption prior to emission. The prime on the first sum symbol in Eq. (8.15) indicates that the term $l = i$, $j = k$, $r = q$, $(\pm)_r = (\pm)_q$ describing the direct return to the initial state ψ_0 is excluded.

The order of terms in Eq. (8.16) corresponds to the order of the transitions shown in Fig. 32. The electron interacting first is denoted by i, the emitted or absorbed photon is q. The second process may be the inverse electron transition by interaction with a second photon r, the inverse photon transition by interaction with a second electron j, or an interaction of a second electron j with a second photon r. The two remaining steps have to bring back the electrons and the photons to their initial states. There exist two possibilities in each of the cases mentioned. Figure 32 shows two additional possibilities for this return in the latter case, where two different electrons i and j interact with two different photons q and r. These additional processes do not represent a photon-exchange, each electron returns to its initial state by interaction with the same photon. The resulting energy terms can be cancelled against those arising by expanding the second order energy expression (8.12) with respect to $E - E_0$.

Let us now arrange the terms in Eq. (8.16) in a different way. In some processes electrons i and j are brought to the states k and l by interaction with the same photon, while the second photon causes the electron transitions back to states i and j. In other cases electrons i and j are brought to states k and l by interaction with different photons, while the inverse electron transitions are caused by the interchanged photons. By interchanging the denotation of the electrons and the order of emission and absorption in the former terms, and the denotation of the electrons and that of the photons in the latter terms, we find that all bilinear terms in the photon occupation numbers $n_q + \frac{1}{2}$, $n_r + \frac{1}{2}$ vanish. After a tedious but elementary transformation of the partial fractions appearing in Eq. (8.16) we obtain

$$
\begin{aligned}
E_4 = -\tfrac{1}{2} \sum_{q,\pm} \sum_{r,\pm} \sum_{ik} f_i(1 - f_k) \sum_{jl} f_j(1 - f_l) \Bigg\{ &\frac{(\pm)_r(n_q + \frac{1}{2})}{\pm \hbar\omega_q \pm \hbar\omega_r} \\
\cdot \left(\frac{U^\pm(kiq)\,U^\pm(ikr)}{E_k - E_i \pm \hbar\omega_q} + \frac{U^\pm(ikq)\,U^\pm(kir)}{E_k - E_i \mp \hbar\omega_q} \right) & \\
\cdot \left(\frac{U^\mp(ljq)\,U^\mp(jlr)}{E_l - E_j \mp \hbar\omega_q} + \frac{U^\mp(jlq)\,U^\mp(ljr)}{E_l - E_j \pm \hbar\omega_q} \right) & \\
- (\pm)_q(\pm)_r \frac{U^\pm(kiq)\,U^\pm(ikr)}{(E_k - E_i \pm \hbar\omega_q)(E_k - E_i \mp \hbar\omega_r)} & \\
\cdot \left(\frac{U^\mp(ljq)\,U^\mp(jlr)}{E_l - E_j + E_k - E_i} + \frac{U^\mp(jlq)\,U^\mp(ljr)}{E_l - E_j - E_k + E_i} \right) & \Bigg\}.
\end{aligned}
\tag{8.17}
$$

In spite of considering two-photon processes, we still find the energy levels (8.17) to be equidistant with respect to the photon occupation numbers $n_q + \frac{1}{2}$. This suggests that it is permissible to neglect the effect of the electron-photon interaction on statistics and to occupy the electron states and the photon states according to Fermi statistics and to Bose statistics. We shall check this question in Section 8.4 by studying the creation and annihilation operators of the resulting quasi-electrons and quasi-photons.

One might suggest that Eq. (8.17) could be obtained more easily by transforming the Hamiltonian (8.1) from the Schrödinger representation to the interaction representation. Among the terms quadratic in the electron-photon interaction operator one finds two-photon excitations via an intermediate electron transition, and two-electron excitations via a photon exchange. However, on attempting to obtain the first term in Eq. (8.17) from second order perturbation theory with respect to the two-photon excitation operator, it turns out that the matrix elements must be symmetric in the interacting photons q and r, which is not true for the matrix elements appearing in Eq. (8.17).

8.3. Zero Temperature Limit

It is obvious from Eq. (8.17) that the effect of the electrons is basically covered by the quantities

$$
\begin{aligned}
&X(\xi; \pm q, \pm r) \\
&= \sum f_i(1 - f_k)\left(\frac{U^\pm(kiq)\,U^\pm(ikr)}{E_k - E_i + \xi} + \frac{U^\pm(ikq)\,U^\pm(kir)}{E_k - E_i - \xi} \right).
\end{aligned}
\tag{8.18}
$$

$X(\xi; \pm q, \pm r)$ replaces the susceptibilities introduced in the preceding sections. In contrast to the quantities used before, it explicitly contains the position of the electron orbitals involved, i.e. $X(\xi; \pm q, \pm r)$ covers the spatial variation of the photon field as well.

When we attempt to simplify expression (8.17) by means of complex integration techniques, we note the characteristic structure of the energy denominators. The energy change of one of the interacting electrons and photons appears three times, that of the others just once. By associating these denominators with a residue caused by the energy change of the distinguished transition, it is possible to split all terms into independent contributions of the interacting electrons and photons. We obtain the general expression

$$E_4 = (8\pi i)^{-1} \int_C d\xi \sum_{q\pm} \sum_{r\pm} \frac{(\pm)_q}{\xi \mp \hbar\omega_q} X(\xi; \pm q, \pm r) \frac{(\pm)_r}{\xi \pm \hbar\omega_r} X(-\xi; \mp q, \mp r).$$

(8.19)

The contour C of integration in Eq. (8.19) encloses once all electronic poles $\xi + E_k - E_i = 0$, and $2n_q + 1$ times all photonic poles $\xi + \hbar\omega_q = 0$ on the negative real axis. This is generally a very inconvenient contour of integration. However, in the zero temperature limit, when our main interest is in the energy gain of the ground state $n_q = 0$, the contour has to enclose all poles once and may be chosen to run along the imaginary axis from $-i\infty$ to $+i\infty$. This is the favorable contour of integration constantly found in all preceding investigations.

To obtain the dispersion energy between two particles 1 and 2, we split the total electronic susceptibility $X(\xi; \pm q, \pm r)$ into those of the electrons localized at particle 1 and particle 2,

$$X_j(\xi; \pm q, \pm r)$$
$$= \sum_{ik\in j} f_i(1 - f_k)\left(\frac{U^\pm(kiq)U^\pm(ikr)}{E_k - E_i + \xi} + \frac{U^\pm(ikq)U^\pm(kir)}{E_k - E_i - \xi}\right).$$

(8.20)

Then, again interchanging the notation of photons q and r and of emission and absorption we find

$$\Delta E_{12} = -(4\pi i)^{-1} \int_{-i\infty}^{+i\infty} d\xi \sum_{q\pm} \sum_{r\pm} \frac{X_1(\xi; \pm q, \pm r)}{\hbar\omega_q \mp \xi} \frac{X_2(-\xi; \mp q, \mp r)}{\hbar\omega_r \pm \xi}.$$ (8.21)

Expression (8.21) for the dispersion energy in the zero temperature limit is obviously in agreement with the findings obtained by the semiclassical approaches. The susceptibilities $X_j(\xi; \pm q, \pm r)$ cover the properties of the electrons and of the photons as well. The product $X_1(\xi; \pm q, \pm r)$ $\cdot X_2(-\xi; \mp q, \mp r)$ contains the product $U^\pm(kiq)U^\mp(ljq)$ of the electron-photon interaction parameters at different positions of emission and absorption, and thus depends on the separation of particles 1 and 2 in the same way as the transposition parameters $V_{mn}^{\mu}(Kr_{21})$, $W_{mn}^{\mu}(Kr_{21})$ appearing in the addition theorem (5.36).

The fourth order perturbation energy (8.21) clearly equals the first term of a more general representation of the free energy of attraction according to Eqs. (3.48), (3.49). The respective dispersion function $G(\xi)$ is split into four subdeterminants $G(\xi; \pm, \pm)$ corresponding to the order of emission and absorption of photons q and r,

$$G(\xi; \pm, \pm)$$

$$= \begin{vmatrix} 1 & \dfrac{X_1(\xi; \pm q, \pm r)}{\hbar\omega_q \mp \xi} & \dfrac{X_1(\xi; \pm q, \pm s)}{\hbar\omega_q \mp \xi} & \cdots \\[2ex] \dfrac{X_2(-\xi; \mp q, \mp r)}{\hbar\omega_r \pm \xi} & 1 & \dfrac{X_1(\xi; \pm r, \pm s)}{\hbar\omega_r \mp \xi} & \cdots \\[2ex] \dfrac{X_2(-\xi; \mp q, \mp s)}{\hbar\omega_s \pm \xi} & \dfrac{X_2(-\xi; \mp r, \mp s)}{\hbar\omega_s \pm \xi} & 1 & \cdots \\[2ex] \cdot & \cdot & \cdot & \cdots \end{vmatrix}. \quad (8.22)$$

8.4. Quasi-Particles

With increasing temperature an increasing number of excited states (8.9) become occupied. To find the free energy of interaction at finite temperatures, we have to multiply the excited energy levels (8.10) with their statistic weight and integrate over all states. This raises the question regarding statistics. The energy levels of particles obeying Fermi statistics are occupied according to the Fermi distribution only if the particles are independent, i.e. if the total energy equals the sum over the energies of the individual Fermions. Similarly, we may apply the Bose distribution to the occupation of the energy levels of independent Bosons only. In the present investigations on electron-photon interactions, we obtained energy contributions of order two in $U^{\pm}(kiq)$, which are bilinear in the occupation numbers, therefore we expect deviations from the Fermi and Bose distributions of the same order of magnitude.

We noted that the energy terms (8.8) and (8.12) of order zero and two do not depend on the separation of particles 1 and 2. Accordingly, we calculated the free energy of interaction from the fourth order energy expression (8.16). Considering this term, it is certainly consistent to calculate the average occupation numbers of the excited states from Fermi and Bose statistics. The error is at worst of order six in the interaction parameter.

However, since the statistic weight of the excited states is affected by the electron-photon interaction, there might well arise a free energy of attraction from the energy terms of order zero and two. We have to check whether with decreasing separation there is a redistribution of occupied states, which gives rise to an energy of interaction of order four in $U^{\pm}(kiq)$ as well.

Let us now check whether the need for independent particles is more appropriately met by considering quasi-particles rather than interacting electrons and photons. We noted in Section 8.2 that the energy levels corresponding to the perturbed states (8.9) are equidistant with respect to the photon occupation numbers n_q. This result holds in spite of the fact that the energy terms of order four arise from two-photon exchange interactions, suggesting that at least the quasi-photons represent a system of independent particles.

The perturbed states (8.9) arise from the unperturbed states (8.7) by a one-to-one correspondence, i.e. they are characterized by the occupation numbers f_i and n_q of the electron states i and of the photon states q in the same way as the unperturbed states. Accordingly, we may introduce quasi-electron creation and annihilation operators C_i^+ and C_i and quasi-photon creation and annihilation operators A_q^+ and A_q in a fully analogous manner to the unperturbed case. We require

$$C_i^+ \psi(0) = \psi(1); \qquad C_i \psi(1) = \psi(0) \tag{8.23}$$

$$A_q^+ \psi(n_q) = \psi(n_q + 1); \qquad A_q \psi(n_q) = \psi(n_q - 1) \tag{8.24}$$

where ψ represents a perturbed orbital (8.9) and the argument shown is the occupation number of the electron state i or of the photon state q, respectively.

To derive explicit representations of the quasi-particle operators defined by Eqs. (8.23), (8.24), we express the perturbed states ψ in terms of the unperturbed states ψ_0 according to Eqs. (8.9), (8.11), (8.15). We obtain

$$
\begin{aligned}
C_i^+ &\left\{ 1 + \sum_{jk} \sum_{q\pm} \frac{U^{\pm}(kjq)a_q^{\pm}}{E - E_0 - (E_k - E_j \pm \hbar\omega_q)} c_k^+ c_j + \cdots \right\} \\
&= \left\{ 1 + \sum_{jk} \sum_{q\pm} \frac{U^{\pm}(kjq)a_q^{\pm}}{E - E_0 - (E_k - E_j \pm \hbar\omega_q)} c_k^+ c_j + \cdots \right\} c_i^+
\end{aligned}
\tag{8.25}
$$

and equivalent equations for the remaining quasi-particle operators C_i, A_q^+, A_q.

Restricting ourselves for the present to terms up to order two in the interaction parameters $U^{\pm}(ljr)$, we cancel the terms $E - E_0$ in the denominators of Eq. (8.25). Iteratively solving for C_i^+, we obtain

$$C_i^+ = c_i^+ - \sum_k \sum_{q\pm} \frac{U^\pm(kiq)a_q^\pm}{E_k - E_i \pm \hbar\omega_q} c_k^+ (c_i c_i^+ - c_i^+ c_i)$$

$$+ \sum_{jl} \sum_{r\pm}' \sum_k \sum_{q\pm} \frac{U^\pm(ljr)}{E_l - E_j \pm \hbar\omega_r + E_k - E_i \pm \hbar\omega_q} \frac{U^\pm(kiq)}{E_k - E_i \pm \hbar\omega_q} \qquad (8.26)$$

$$\cdot [c_i^+ c_j a_r^\pm, c_k^+(c_i c_i^+ - c_i^+ c_i)a_q^\pm] - c_i^+ \sum_k \sum_{q\pm} \frac{U^\pm(kiq)U^\mp(ikq)}{(E_k - E_i \pm \hbar\omega_q)^2} a_q^\pm a_q^\mp c_k^+ c_k$$

$$C_i = c_i - \sum_k \sum_{q\pm} \frac{U^\pm(ikq)a_q^\pm}{E_i - E_k \pm \hbar\omega_q} (c_i^+ c_i - c_i c_i^+)c_k$$

$$+ \sum_{jl} \sum_{r\pm}' \sum_k \sum_{q=} \frac{U^\pm(ljr)}{E_l - E_j \pm \hbar\omega_r + E_i - E_k \pm \hbar\omega_q} \frac{U^\pm(ikq)}{E_i - E_k \pm \hbar\omega_q} \qquad (8.27)$$

$$\cdot [c_i^+ c_j a_r^\pm, (c_i^+ c_i - c_i c_i^+)c_k a_q^\pm] - c_i \sum_k \sum_{q\pm} \frac{U^\pm(ikq)U^\mp(kiq)}{(E_i - E_k \pm \hbar\omega_q)^2} a_q^\pm a_q^\mp c_k c_k^+$$

and, similarly

$$A_q^+ = a_q^+ - \sum_{ik} \frac{U^-(kiq)}{E_k - E_i - \hbar\omega_q} c_k^+ c_i$$

$$+ \sum_{ik} \sum_{jl}' \sum_{r\pm} \frac{U^\pm(ljr)a_r^\pm}{E_l - E_j \pm \hbar\omega_r + E_k - E_i - \hbar\omega_q} \frac{U^-(kiq)}{E_k - E_i - \hbar\omega_q} [c_i^+ c_j, c_k^+ c_i]$$

$$+ a_q^+ \sum_{ik} \frac{U^+(ikq)U^-(kiq)}{(E_k - E_i - \hbar\omega_q)^2} c_k^+ c_i c_i^+ c_k \qquad (8.28)$$

$$A_q = a_q + \sum_{ik} \frac{U^+(kiq)}{E_k - E_i + \hbar\omega_q} c_k^+ c_i \qquad (8.29)$$

$$+ \sum_{ik} \sum_{jl}' \sum_{r\pm} \frac{U^\pm(ljr)a_r^\pm}{E_l - E_j \pm \hbar\omega_r + E_k - E_i + \hbar\omega_q} \frac{U^+(kiq)}{E_k - E_i + \hbar\omega_q} [c_k^+ c_i, c_l^+ c_j]$$

$$- a_q \sum_{ik} \frac{U^+(kiq)U^-(ikq)}{(E_k - E_i + \hbar\omega_q)^2} c_k^+ c_i c_i^+ c_k .$$

Each electron creation or annihilation in state i is coupled to electron creations and annihilations in other states k via photon emission and absorption. The operators C_i^+, C_i create and annihilate quasi-electrons. Similarly, each photon emission or absorption in state q is coupled to an electron transition from state i to state k. The operators A_q^+, A_q emit and absorb quasi-photons. The brackets in Eqs. (8.26)–(8.29) denote the commutators. We distinguish two different terms of order two in the interaction parameters. The second order processes leading to the same final situation as the direct process are excluded by the prime on the sum symbol in the first term and give rise to the second term instead.

By substituting the quasi-particle operators (8.26)–(8.29) into the Hamiltonian (8.1), we obtain

$$H = \sum_i E_i C_i^+ C_i + \sum_q \tfrac{1}{2} \hbar \omega_q (A_q A_q^+ + A_q^+ A_q) \tag{8.30}$$

$$- \sum_{ikq} \left(\frac{U^+(kiq)\, U^-(ikq)}{E_k - E_i + \hbar \omega_q} A_q A_q^+ + \frac{U^-(kiq)\, U^+(ikq)}{E_k - E_i - \hbar \omega_q} A_q^+ A_q \right) C_i^+ C_k C_k^+ C_i.$$

The Hamiltonian is now in its diagonal form. The quasi-particle operators describe the transitions between the exact eigenstates of the electron-photon system under investigation. The terms quadratic in $U^{\pm}(kiq)$ do not really couple the eigenstates, but modify the eigenvalues. They yield the second order interaction energy (8.12). The terms of order four in $U^{\pm}(kiq)$ are obtained in a similar manner by substituting quasi-particle creation and annihilation operators for the occupation numbers in the fourth order energy expression (8.17).

Symmetrizing the products $A_q A_q^+$ and $A_q A_q^+$ of the Boson creation and annihilation operators by means of the commutation relation (8.4) we obtain alternatively.

$$H = \sum_i E_i C_i^+ C_i$$

$$- \tfrac{1}{2} \sum_{ik} \sum_q \left(\frac{U^+(kiq)\, U^-(ikq)}{E_k - E_i + \hbar \omega_q} - \frac{U^-(kiq)\, U^+(ikq)}{E_k - E_i - \hbar \omega_q} \right) C_i^+ C_k C_k^+ C_i \tag{8.31}$$

$$+ \sum_q \tfrac{1}{2} (A_q A_q^+ + A_q^+ A_q)$$

$$\cdot \left[\hbar \omega_q - \sum_{ik} \left(\frac{U^+(kiq)\, U^-(ikq)}{E_k - E_i + \hbar \omega_q} + \frac{U^-(kiq)\, U^+(ikq)}{E_k - E_i - \hbar \omega_q} \right) C_i^+ C_i \right].$$

Each term in Eq. (8.31) contains the creation and annihilation operators of at most two different particles. The energy due to the quasi-photons is additively composed of that of the individual quasi-photons, i.e. the quasi-photons constitute a system of independent Bosons. We may occupy the quasi-photon states according to the Bose distribution.

However, each quasi-photon transition still affects the energy of the quasi-electrons, and each quasi-electron transition still affects the energy of the quasi-photons. We may not simultaneously occupy the quasi-electron states according to the Fermi distribution and quasi-photon states according to the Bose distribution. This is permissible only after renormalization of the quasi-particles considered.

8.5. Gibbs Distribution

Finally, we are interested in the free energy of the electron-photon system under investigation. The free energy of attraction of particles 1 and 2 is the difference between the free energies at separations r_{21} and infinity. The free energy of any system of particles equals

$$F = -kT \ln Z \qquad (8.32)$$

where Z is the partition function weighting all eigenstates according to the Gibbs distribution,

$$Z = \sum_{f_i f_j \dots} \sum_{n_q n_r \dots} \exp[-E(f_i f_j \dots; n_q n_r \dots)/kT]. \qquad (8.33)$$

We have to sum over all possible distributions of the quasi-electrons among the electron states i and over all quasi-photon occupation numbers n_q. Using the fact that the energy levels $E(f_i f_j \dots; n_q n_r \dots)$ are equidistant with respect to the photon occupation numbers inclusive of terms of order four in $U^\pm(kiq)$, we may carry out the summations over $n_q, n_r, \dots$ explicitly and obtain

$$Z = \sum_{f_i f_j \dots} \exp[-E_{el}(f_i f_j \dots)/kT] \prod_q [2 \sinh(\langle \hbar \omega_q \rangle / 2kT)]^{-1}. \qquad (8.34)$$

$\langle \hbar \omega_q \rangle$ is the renormalized photon energy (8.14). It covers all energy terms linear in $n_q + \frac{1}{2}$ and thus depends on the occupation numbers $f_i, f_j, \dots$ of the electron states. According to Eqs. (8.12), (8.17), and (8.18) we find

$$\langle \hbar \omega_q \rangle = \hbar \omega_q - X(\hbar \omega_q; +q, -q)$$

$$- \sum_{r, \pm} \frac{(\pm)_r}{\hbar \omega_q \pm \hbar \omega_r} X(\hbar \omega_q; +q, \pm r) X(-\hbar \omega_q; -q, \mp r). \qquad (8.35)$$

To carry out the summation in Eq. (8.34) over all electron states, it is necessary to expand the photon terms with respect to $\langle \hbar \omega_q \rangle - \hbar \omega_q$. We obtain all terms up to order four in the interaction parameters $U^\pm(kiq)$ by expanding up to terms quadratic in $\langle \hbar \omega_q \rangle - \hbar \omega_q$. However, bearing in mind that all terms which do not depend on the separation of particles 1 and 2 cancel when the free energy of attraction is calculated, it is sufficient to indicate the terms quadratic in $\langle \hbar \omega_q \rangle - \hbar \omega_q$ by dots. Hence,

$$Z = \prod_q [2 \sinh(\hbar \omega_q/2kT)]^{-1} \sum_{f_i f_j \dots} \exp[-E_{el}(f_i f_j \dots)/kT]$$

$$\cdot \left[1 - \sum_q (\langle \hbar \omega_q \rangle - \hbar \omega_q)(2kT)^{-1} \coth(\hbar \omega_q/2kT) + \cdots \right]. \qquad (8.36)$$

By including the photon renormalization term in the exponential we obtain

$$Z = \prod_q [2\sinh(\hbar\omega_q/2kT)]^{-1} \sum_{f_i f_j \cdots} \exp[-\langle E_{el}(f_i f_j \cdots)\rangle/kT] \qquad (8.37)$$

where

$$\langle E_{el}(f_i f_j \cdots)\rangle = E_{el}(f_i f_j \cdots) + \sum_q \tfrac{1}{2}[\langle\hbar\omega_q\rangle - \hbar\omega_q]\coth(\hbar\omega_q/2kT) + \cdots . \qquad (8.38)$$

The renormalization of the quasi-photon energies entails a renormalization of the quasi-electron energies. The renormalized quasi-electron Hamiltonian results from the total Hamiltonian (8.31) by replacing the photon occupation numbers $n_q + \tfrac{1}{2}$ by $\tfrac{1}{2}\coth(\hbar\omega_q/2kT)$. These are just the average occupation numbers given by the Bose distribution (3.6). $\langle E_{el}(f_i f_j \cdots)\rangle$ is obtained from Eqs. (8.10), (8.12), (8.17) through the same substitution.

The quasi-electron transitions, in contrast to the quasi-photons, are strongly interdependent. The energy necessary for creation or annihilation of an electron in state i depends explicitly on the occupation of the remaining electron states $j, k, \ldots$. The summation over a particular electron state i in the partition function (8.37) entails a renormalization of all remaining electron states $j, k \ldots$. Expanding

$$Z_{el} = \sum_{f_i f_j \cdots} \exp[-\langle E_{el}(f_i f_j \cdots)\rangle/kT] \qquad (8.39)$$

with respect to the change in creation energy $\langle E_{el}(1 f_j \cdots) - E_{el}(0 f_j \cdots) - E_i\rangle$ in state i, we obtain, in a similar manner as when summing over the photon states

$$Z_{el} = \sum_{f_j \cdots} \{\exp[-\langle E_{el}(0 f_j \cdots)\rangle/kT] + \exp[-\langle E_{el}(1 f_j \cdots)\rangle/kT]\} \qquad (8.40)$$

$$Z_{el} = \sum_{f_j \cdots} \exp[-\langle E_{el}(0 f_j \cdots)\rangle/kT]\{1 + \exp[-E_i/kT]$$
$$\cdot (1 - \langle E_{el}(1 f_j \cdots) - E_{el}(0 f_j \cdots) - E_i\rangle/kT + \cdots)\} \qquad (8.41)$$

and

$$Z_{el} = \{1 + \exp[-E_i/kT]\} \sum_{f_j \cdots} \exp[-(\langle E_{el}(0 f_j \cdots)\rangle$$
$$+ \langle f_i\rangle\langle E_{el}(1 f_j \cdots) - E_{el}(0 f_j \cdots) - E_i\rangle)/kT] \qquad (8.42)$$

where $\langle f_i\rangle$ equals the Fermi distribution

$$\langle f_i\rangle = \{1 + \exp[-E_i/kT]\}^{-1} . \qquad (8.43)$$

Carrying out the summation over state i in the partition function (8.39) we obtain the Fermi distribution (8.43). The energy levels corresponding to the remaining states $j, k, \ldots$ are derived by substituting $\langle f_i \rangle$ for f_i in Eqs. (8.10), (8.12), and (8.17). Successively summing over all remaining states $j, k, \ldots$, we find

$$Z_{el} = \prod_k \{1 + \exp[-E_k/kT]\} \exp[-\langle E_{el}(\langle f_i \rangle \langle f_j \rangle \ldots)\rangle/kT]. \qquad (8.44)$$

All electron states are on the average occupied according to the Fermi distribution (8.43). All photon states are on the average occupied according to the Bose distribution (3.6). On substituting Eq. (8.44) in Eqs. (8.37) and (8.32) we find for the free energy of attraction between particles 1 and 2

$$\Delta E_{12} = [F]_\infty^{r21} = [\langle E_{el}(\langle f_i \rangle \langle f_j \rangle \ldots)\rangle]_\infty^{r21}. \qquad (8.45)$$

The free energy of attraction between particles 1 and 2 results by substituting the Fermi distribution and the Bose distribution into the fourth order energy expression (8.17).

8.6. Free Energy of Attraction

The proposed successive summation over all electron states i involves the cancellation of terms quadratic in the electron-photon interaction parameter $U^\pm(kiq)$ each time, whereas the terms of the fourth order which depend on the separation of particles 1 and 2 are retained. This procedure incures the unnecessary risk of omitting terms quadratic in $U^\pm(kiq)$ in the electron and photon distribution functions of the separated particles 1 and 2. In order to obtain the energy of attraction correctly up to terms of the sixth order in the interaction parameters we now sum the partition function (8.33) more rigorously.

The energy levels according to Eqs. (8.10), (8.12), (8.17) can be clearly split into a contribution $U_1(f_i f_k \ldots)$ of the electrons located at particle 1, a contribution $U_2(f_j f_l \ldots)$ of the electrons located at particle 2, and a mixed contribution which is additively composed from products of susceptibilities referring to the two particles

$$E = U_1(f_i f_k \ldots) + U_2(f_j f_l \ldots) + \sum_n V_{1n}(f_i f_k \ldots) V_{2n}(f_j f_l \ldots). \qquad (8.46)$$

Equation (8.46) holds with respect to all orders in the electron-photon interaction parameter. Substituting Eq. (8.46) into (8.39), we expand the exponential term with respect to the mixed contribution $\sum V_{1n} V_{2n}$, which starts with terms of the fourth order in the interaction parameter. We obtain

$$Z_{el} = \sum_{f_i f_k \cdots} \sum_{f_j f_l \cdots} \exp[-(U_1 + U_2)/kT] \left(1 - \sum_n V_{1n} V_{2n}/kT\right) \qquad (8.47)$$

and

$$Z_{el} = \sum_{f_i f_k \cdots} \exp[-U_1(f_i f_k \cdots)/kT] \sum_{f_j f_l \cdots} \exp[-U_2(f_j f_l \cdots)/kT]$$
$$\cdot \left(1 - \sum_n \langle V_{1n} \rangle \langle V_{2n} \rangle /kT\right) \qquad (8.48)$$

where $\langle V_{jn} \rangle$ results by averaging the susceptibility $V_{jn}(f_i f_k \cdots)$ of particle j according to the Gibbs distribution at that particle

$$\langle V_{jn} \rangle = \frac{\sum\limits_{f_i f_k \cdots} V_{jn}(f_i f_k \cdots) \exp[-U_j(f_i f_k \cdots)/kT]}{\sum\limits_{f_i f_k \cdots} \exp[-U_j(f_i f_k \cdots)/kT]}. \qquad (8.49)$$

Hence,

$$\Delta E_{12} = \sum_n \langle V_{1n} \rangle \langle V_{2n} \rangle. \qquad (8.50)$$

We obtain the free energy of attraction between particles 1 and 2 by occupying the quasi-electron states of the single particles 1 and 2 according to the Gibbs distribution.

The Gibbs distribution (8.49) differs from the Fermi distribution (8.43), if the total energy $U_j(f_i f_k \cdots)$ of the electrons at particle j is not additively composed of the energy of the single electrons. The lowest order term violating this additivity arises in E_2 according to Eq. (8.12). The term corresponding to the spontaneous emission of a photon q, i.e. the term being independent of n_q, is bilinear in $f_i f_k$. After renormalization with respect to this term, we are left with true deviations of the Fermi distribution from the Gibbs distribution of the fourth order in $U^{\pm}(kiq)$.

Equation (8.50) holds inclusive of terms of the sixth order in $U^{\pm}(kiq)$, i.e. by expanding the free energy with respect to $\Sigma V_{1n} V_{2n}$, we neglected terms of the eighth order. It would now be consistent to extend the calculation of the energy levels (8.10) to terms E_6 of the sixth order in the interaction parameter.

Applying Eq. (8.50) to the fourth order energy term (8.17), we find that the state density integration technique used in Section 8.3 is even more powerful than in the zero temperature limit. Substitution of the Bose distribution $\langle n_q + \frac{1}{2} \rangle$ and the Fermi distribution $\langle f_i \rangle$ for the particular occupation numbers $n_q + \frac{1}{2}$ and f_i implies that the contour C of integration in Eq. (8.19) has to enclose all photon poles $\coth(\hbar \omega_q/2kT)$ times. We may account for this weight of the photon poles by multiplying the integrand in Eq. (8.19) with the weight factor $\coth(-\xi/2kT)$. This means that all electron poles $\xi + E_k - E_i = 0$ obtain the corresponding weight $\coth((E_k - E_i)/2kT)$. Rewriting this term as

$$\coth(E_k - E_i)/2kT = \frac{\langle f_i \rangle \langle 1 - f_k \rangle + \langle f_k \rangle \langle 1 - f_i \rangle}{\langle f_i \rangle \langle 1 - f_k \rangle - \langle f_k \rangle \langle 1 - f_i \rangle} \tag{8.51}$$

we find that the denominator in Eq. (8.51) cancels with the respective term in the susceptibility $X(\xi; \pm q, \pm r)$. Hence,

$$E_4 = (8\pi i)^{-1} \int_C d\xi \coth(\xi/2kT) \sum_{q\pm} \sum_{r\pm} \frac{X(\xi; \pm q, \pm r)}{\hbar\omega_q \mp \xi} \frac{X(-\xi; \mp q, \mp r)}{\hbar\omega_r \pm \xi}$$

$$\tag{8.52}$$

where the contour of integration in Eq. (8.52) encloses either the positive or the negative real axis. Shifting the contour to the imaginary axis in a similar manner to our semiclassical procedure in Section 3.4, we find the free energy of attraction

$$\Delta E_{12} = -(4\pi i)^{-1} \int_{-i\infty}^{+i\infty} d\xi \coth(\xi/2kT)$$

$$\cdot \sum_{q\pm} \sum_{r\pm} \frac{X_1(\xi; \pm q, \pm r)}{\hbar\omega_q \mp \xi} \frac{X_2(-\xi; \mp q, \mp r)}{\hbar\omega_r \pm \xi} \tag{8.53}$$

and

$$\Delta E_{12} = (4\pi i)^{-1} \int_{-i\infty}^{+i\infty} d\xi \coth(\xi/2kT) \sum_{\pm, \pm} \ln G(\xi; \pm, \pm) \tag{8.54}$$

where $G(\xi; \pm, \pm)$ is given by Eq. (8.22). The contour of integration in Eqs. (8.53), (8.54) by-passes the poles of $\coth(\xi/2kT)$ on the right-hand side. Since the principal values of these integrals vanish, we obtain the alternative expression

$$\Delta E_{12} = \tfrac{1}{2} kT \sum_{n=-\infty}^{+\infty} \sum_{\pm, \pm} \ln G(2\pi nkT; \pm, \pm). \tag{8.55}$$

The final Eqs. (8.53)–(8.55) for the free energy of attraction between particles 1 and 2 confirm and refine the semiclassical results obtained in Chapter 3–6. There are contributions from the photons and the electrons to the free energy of attraction as well. The factor $\coth(\xi/2kT)$ is not only caused by the Bose distribution, but is also required by the Fermi distribution. In accordance with the symmetric use of retarded and advanced susceptibilities in the case of damping in Section 3.6, we now distinguish between emission of a photon prior to absorption and absorption of a photon prior to emission. The result is a symmetric final integration over the dispersion function along the full imaginary frequency axis in both cases.

Acknowledgement. My first activities in the field of van der Waals attraction were stimulated by the experimental investigations on adhesion conducted by H. Krupp and G. Walter at Battelle-Institut, Frankfurt/Main. Thereafter I greatly enjoyed the cooperation with D. Bargeman and F. van Voorst Vader of the Unilever Research Laboratory, Vlaardingen. The main task for the theoretician among experimentalists was to interpret the effect of different shapes and susceptibilities on the attraction of particles at close distance. It was Adrian Parsegian, National Institute of Health, Washington, who pointed out to me the numerous applications of the theory in biology and suggested to include also cylindrical shapes in the investigation. In addition, I had clarifying discussions with K. Schram, Rijksuniversiteit Utrecht, on the influence of damping and with E. Gerlach, now at Technische Hochschule Aachen, on the contribution of surface modes. An excellent opportunity of reconsidering all concepts of van der Waals attraction arose when I accepted a Visiting Fellowship at the Institute of Advanced Studies, Australian National University, Canberra. A long series of seminars and colloquia and many most stimulating discussions with Barry Ninham and his co-workers gave rise to a critical review and numerous extensions of the theory. It was at that time that the first draft of the present monograph was written. Several sections which still appeared insufficiently founded were revised and tested a second time in a lecture course at Frankfurt University. I am happy to thank my wife Walburga, my children and my friends for their understanding and help during the preparation of the final version. I also extend my thanks to Ingrid Voigt-Martin for carefully checking the English draft and Gabriele Klempert for typing the manuscript.

Bibliography

The following references are arranged according to their main topic. Most of them could have been placed under the preceding or following topics as well.

Mathematical Functions

1. Abramowitz, M., Stegun, I. A.: Handbook of Mathematical Functions. New York: Dover 1965.
2. Erdélyi, A., Magnus, W., Oberhettinger, F., Tricomi, F. G.: Higher transcendental functions, Vol. 1. New York: McGraw-Hill 1953.
3. Erdélyi, A., Magnus, W., Oberhettinger, F., Tricomi, F. G.: Higher transcendental functions, Vol. 2. New York: McGraw-Hill 1953.
4. Morse, P. M., Feshbach, H.: Methods of theoretical physics, Part II, Section 13.1. New York: McGraw-Hill 1953.

Reviews

5. Margenau, H.: Rev. Mod. Phys. **11**, 1 (1939).
6. Hirschfelder, J. O., Curtis, C. R., Bird, R. B.: Molecular theory of gases and liquids. New York: Wiley 1954.
7. Landau, L. D., Lifshitz, E. M.: Statistical Physics, Chapter 12. London-Paris: Pergamon 1959.
8. Pitzer, K. S.: Advan. Chem. Phys. **2**, 59 (1959).
9. Abrikosov, A. A., Gorkov, L. P., Dzyaloshinski, I. E.: Methods of quantum field theory in statistical physics, Chapter 6. Englewood Cliffs: Prentice-Hall 1963.
10. Sinanoglu, O.: Advan. Chem. Phys. **6**, 315 (1964).
11. Dalgarno, A., Davison, W. D.: Advan. At. Molec. Phys. **2**, 1 (1966).
12. Hirschfelder, J. O., Meath, W. J.: Advan. Chem. Phys. **12**, 3 (1967).

13. Buckingham, A. D.: Advan. Chem. Phys. **12**, 107 (1967).
14. Dalgarno, A.: Advan. Chem. Phys. **12**, 143 (1967).
15. Power, E. A.: Advan. Chem. Phys. **12**, 167 (1967).
16. Linder, B.: Advan. Chem. Phys. **12**, 225 (1967).
17. Sinanoglu, O.: Advan. Chem. Phys. **12**, 283 (1967).
18. Krupp, H.: Advan. Colloid Interface Sci. **1**, 111 (1967).

Orientation, Induction, and Dispersion Forces

19. Keesom, W. H.: Koninkl. Ned. Akad. Wetenschap. Proc., **18**, 636 (1916); **23**, 939 (1920); **24**, 162 (1921).
20. Keesom, W. H.: Z. Physik **22**, 129, 643 (1921); **23**, 225 (1922).
21. Debye, P.: Z. Physik **21**, 178 (1920); **22**, 302 (1921).
22. Eisenschitz, R., London, F.: Z. Physik **60**, 491 (1930).
23. London, F.: Z. Physik **63**, 245 (1930).
24. London, F.: Z. Phys. Chem. **B11**, 222 (1930).
25. London, F.: Trans. Faraday Soc. **33**, 8 (1937).

Retardation

26. Casimir, H. B. G., Polder, D.: Phys. Rev. **73**, 360 (1948).
27. Casimir, H. B. G.: Koninkl. Ned. Akad. Wetenschap. Proc. **51**, 793 (1948).
28. Lifshitz, E. M.: Zh. Eksp. Teor. Fiz. **29**, 94 (1955) (Soviet Phys. JETP **2**, 73 (1956)).
29. Dzyaloshinskii, I. E., Lifshitz, E. M., Pitaevskii, L. P.: Zh. Eksp. Teor. Fiz. **37**, 229 (1959) (Soviet Phys. JETP **10**, 161 (1960)).
30. Aub, M. R., Zienau, S.: Proc. Roy. Soc. (Lond.) A**257**, 464 (1960).
31. Gann, V. V.: Zh. Eksp. Teor. Fiz. **54**, 994 (1968) (Soviet Phys. JETP **27**, 529 (1968)).
32. Feinberg, G., Sucher, J.: Phys. Rev. A**2**, 2395 (1970).
33. Langbein, D.: Phys. Rev. B**2**, 3371 (1970).

State Density Integration

34. Mahan, G. D.: J. Chem. Phys. **43**, 1569 (1965).
35. van Kampen, N. G., Nijboer, B. R. A., Schram, K.: Phys. Letters **26** A, 307 (1968).
36. Ninham, B. W., Parsegian, V. A., Weiss, G. H.: J. Stat. Phys. **2**, 323 (1970).
37. Richmond, P., Ninham, B. W.: J. Phys. C: Solid State Phys. **4**, 1988 (1971).
38. Gerlach, E.: Phys. Rev. B**4**, 393 (1971).
39. Gerlach, E.: J. Vac. Sci. Techn. **9**, 747 (1971).
40. Mitchell, D. J., Ninham, B. W., Richmond, P.: Australian J. Phys. **25**, 33 (1972).
41. Mahanty, J., Ninham, B. W.: J. Phys. A: Gen. Phys. **5**, 1447 (1972).

Multiplet Interactions

42. Axilrod, B. M., Teller, E.: J. Chem. Phys. **11**, 299 (1943).
43. Axilrod, B. M.: J. Chem. Phys. **19**, 719 (1951).
44. Jansen, L., McGinnes, R. T.: Phys. Rev. **104**, 961 (1956).
45. Bade, W. L.: J. Chem. Phys. **27**, 1280 (1957).
46. Bade, W. L., Kirkwood, J. G.: J. Chem. Phys. **27**, 1284 (1957).
47. Jansen, L.: Phys. Rev. **125**, 1798 (1962).
48. Renne, M. J., Nijboer, B. R. A.: Chem. Phys. Letters **1**, 317 (1967).
49. Nijboer, B. R. A., Renne, M. J.: Chem. Phys. Letters **2**, 35 (1968).
50. Bell, R. J.: J. Phys. B: Atom. Molec. Phys. **3**, 751 (1970).
51. Langbein, D.: J. Phys. Chem. Solids **32**, 133 (1971).
52. Langbein, D.: J. Phys. A: Gen. Phys. **4**, 471 (1971).
53. Malvien, J. P.: Int. J. Quant. Chem. **5**, 435, 455 (1971).

Macroscopic Approaches

54. Doniach, S.: Phil. Mag. **8**, 129 (1963).
55. Lucas, A.: Physica **35**, 353 (1967).
56. Ruppin, R., Englman, R.: J. Phys. C: Solid State Phys. **2**, 630 (1968).
57. Langbein, D.: J. Phys. A: Math., Nucl. Gen. **6**, 1149 (1973).
58. Dissado, L. A.: J. Phys. C: Solid State Phys. **6**, 158 (1973).

Reaction Fields

59. Linder, B.: J. Chem. Phys. **33**, 668 (1960).
60. Linder, B.: J. Chem. Phys. **37**, 963 (1962).
61. McLachlan, A. D.: Proc. Roy. Soc. A **271**, 387 (1963).
62. McLachlan, A. D.: Proc. Roy. Soc. A **274**, 80 (1963).
63. Linder, B., Hoernschemeyer, D.: J. Chem. Phys. **40**, 622 (1964).
64. Linder, B.: J. Chem. Phys. **40**, 2003 (1964).

Fluctuations

65. Callen, H. B., Welton, T.: Phys. Rev. **83**, 34 (1951).
66. Kershaw, D.: Phys. Rev. B **136**, 1850 (1964).
67. Nelson, E.: Phys. Rev. **150**, 1079 (1966).
68. Boyer, T. H.: Phys. Rev. **174**, 1631 (1968).
69. Boyer, T. H.: Phys. Rev. **174**, 1764 (1968).
70. Boyer, T. H.: Phys. Rev. **182**, 1374 (1969).
71. Boyer, T. H.: Phys. Rev. **186**, 1304 (1969).

Susceptibilities

72. Hopfield, J. J.: Phys. Rev. **112**, 1555 (1958).
73. Kaneyoshi, T.: Progr. Theor. Phys. **41**, 577 (1969).
74. Kats, A. V., Maslov, V. V.: Zh. Eksp. Teor. Fiz. **62**, 496 (1972) (Soviet Phys. JETP **35**, 264 (1972)).
75. Bykov, V. P.: Zh. Eksp. Teor. Fiz. **62**, 505 (1972) (Soviet Phys. JETP **35**, 269 (1972)).
76. Allison, D. C. S., Burke, P. G., Robb, W. D.: J. Phys. B: Atom. Molec. Phys. **5**, 55 (1972).
77. Johnson, P. B., Christy, R. W.: Phys. Rev. B **6**, 4370 (1972).
78. Swain, S.: J. Phys. A: Math. Nucl. Gen. **6**, 192 (1973).
79. Hall, G. G.: Int. J. Quant. Chem. **7**, 15 (1973).

Dielectric Anisotropy

80. Okano, K.: J. Phys. Soc. Japan **20**, 2085 (1965).
81. Imura, H., Okano, K.: Progr. Polymer Phys. Japan **14**, 675 (1971).
82. Parsegian, V. A., Weiss, G. H.: J. Colloid Interface Sci. **40**, 35 (1972).
83. Parsegian, V. A., Weiss, G. H.: J. Adhesion **3**, 259 (1972).
84. Barouch, E., J. W. Perram, Smith, E. R.: Stud. Appl. Math. **52**, 175 (1973).

Multilayers

85. Ninham, B. W., Parsegian, V. A.: J. Chem. Phys. **52**, 4578 (1970).
86. Ninham, B. W., Parsegian, V. A.: J. Chem. Phys. **53**, 3398 (1970).
87. Langbein, D.: J. Adhesion **3**, 213 (1972).
88. Israelachvili, J. N.: Proc. Roy. Soc. (Lond.) A **331**, 39 (1972).
89. Langbein, D.: J. Adhesion **6**, 1 (1974).

Effects of Shape

90. de Boer, J. H.: Trans. Faraday Soc. **32**, 10 (1936).
91. Hamaker, H. C.: Physica **4**, 1058 (1937).
92. Salem, L.: J. Chem. Phys. **37**, 2100 (1962).
93. Zwanzig, R.: J. Chem. Phys. **39**, 2251 (1963).
94. Langbein, D.: J. Adhesion **1**, 237 (1969).
95. Langbein, D.: J. Phys. Chem. Solids **32**, 1657 (1971).
96. Renne, M. J., Nijboer, B. R. A.: Chem. Phys. Letters **6**, 601 (1970).
97. Mitchell, D. J., Ninham, B. W.: J. Chem. Phys. **56**, 1117 (1972).
98. Parsegian, V. A.: J. Chem. Phys. **56**, 4393 (1972).
99. Langbein, D.: Phys. Kondens. Materie **15**, 61 (1972).
100. Mitchell, D. J., Ninham, B. W., Richmond, P.: J. Theor. Biol. **37**, 251 (1972).
101. Mitchell, D. J., Ninham, B. W., Richmond, P.: Biophys. J. **73**, 359 (1973).
102. Mitchell, D. J., Ninham, B. W., Richmond, P.: Biophys. J. **73**, 370 (1973).
103. Langbein, D.: Festkörperprobleme **13**, 85 (1973).
104. Mahanty, J., Ninham, B. W.: J. Phys. A: Gen. Phys. **6**, 1140 (1973).

Dissipation

105. Nijboer, B. R. A., Renne, M. J.: Physica Norvegica **5**, 243 (1971).
106. Davies, B.: Phys. Letters **37** A, 391 (1971).
107. Langbein, D.: Solid State Commun. **12**, 853 (1973).
108. Langbein, D.: J. Chem. Phys. **58**, 4476 (1973).
109. Schram, K.: Phys. Letters **43** A, 282 (1973).
110. Barash Yu. S.: Radiophysica (USSR) **16**, 1086 (1973).

Temperature and Statistics

111. Abrikosov, A. A., Gorkov, L. P., Dzyaloshinskii, I. E.: Zh. Eksp. Teor. Fiz. **36**, 900 (1959) (Soviet Phys. JETP **36**, 636 (1959)).
112. Fierz, M.: Helv. Phys. Acta **33**, 855 (1960).
113. Rosenkrans, J. P., Linder, B., Kromhout, R. A.: J. Chem. Phys. **49**, 2927 (1968).
114. Kromhout, R. A., Linder, B.: J. Chem. Phys. **49**, 1819 (1968).
115. Linder, B., Kromhout, R. A.: J. Chem. Phys. **49**, 1823 (1968).
116. Parsegian, V. A., Ninham, B. W.: Biophys. J. **10**, 664 (1970).

Spatial Dispersion

117. Hargreaves, C. M.: Koninkl. Ned. Akad. Wetenschap. Proc. B **68**, 231 (1965).
118. Gerlach, E.: Chem. Phys. Letters **3**, 669 (1969).
119. Chang, D. B., Cooper, R. L., Drummond, J. E., Young, A. C.: Phys. Letters **37** A, 311 (1971).
120. Davies, B., Ninham, B. W.: J. Chem. Phys. **56**, 5797 (1972).
121. Richmond, P., Davies, B., Ninham, B. W.: Phys. Letters **39** A, 301 (1972).
122. Davies, B., Ninham, B. W., Richmond, P.: J. Chem. Phys. **58**, 744 (1973).
123. Chang, D. B., Cooper, R. L., Drummond, J. E., Young, A. C.: J. Chem. Phys. **59**, 1232 (1973).
124. Heinrichs, J.: Solid State Commun. **13**, 1595 (1973).

Zero Separation

125. Picknett, R. G.: J. Colloid Interface Sci. **29**, 383 (1969).
126. Dahneke, B.: J. Colloid Interface Sci. **40**, 1 (1972).

127. Parsegian, V. A., Ninham, B. W.: J. Colloid Interface Sci. **37**, 332 (1971).
128. Parsegian, V. A., Gingell, D.: J. Adhesion **4**, 283 (1972).
129. Schmit, J., Lucas, A. A.: Sol. State Commun. **11**, 415 (1972).
130. Schmit, J., Lucas, A. A.: Sol. State Commun. **11**, 419 (1972).
131. Barouch, E., Perram, J. W., Smith, E. R.: Chem. Phys. Letters **19**, 131 (1973).
132. Mahanty, J., Ninham, B. W.: J. Chem. Phys. **59**, 6157 (1973).

Adsorption

133. Vold, M. J.: J. Colloid Sci. **16**, 1 (1961).
134. Grimley, T. B.: Proc. Phys. Soc. **79**, 1203 (1962).
135. Grimley, T. B.: Proc. Phys. Soc. **90**, 751 (1967).
136. Grimley, T. B.: Proc. Phys. Soc. **92**, 776 (1967).
137. Grimley, T. B.: J. Am. Chem. Soc. **90**, 3016 (1967).
138. Grimley, T. B., Walker, S. M.: Surface Sci. **14**, 395 (1969).
139. Richmond, P., Sarkies, K. W.: J. Phys. C: Solid State Phys. **6**, 401 (1973).
140. Grimley, T. B., Torrini, M.: J. Phys. C: Solid State Phys. **6**, 868 (1973).
141. Richmond, P., Ninham, B. W., Ottewill, R. H.: J. Colloid Interface Sci. **45**, 69 (1973).

Orbital Overlap

142. Murrell, J. N., Randic, M., Williams, D. R.: Proc. Roy. Soc. A **284**, 566 (1964).
143. Jansen, L.: Phys. Rev. **162**, 63 (1967).
144. Bukta, J. F., Meath, W. J.: J. Quantum Chem. **6**, 1045 (1972).
145. Jansen, L.: Molecular Processes on Solid Surfaces. Ed. E. Drauglis, R. Gretz, R. Jaffee. McGraw-Hill 1969, p. 49.
146. Farberov, D. S., Mitrofanov, V. YA., Men, A. N.: J. Quantum Chem. **6**, 1057 (1972).

Direct Measurements

147. Derjaguin, B. V., Abrikosova, J. J.: Disc. Faraday Soc. **18**, 24 (1954).
148. Derjaguin, B. V., Abrikosova, J. J., Lifshitz, E. M.: Quart. Rev. (London) **10**, 295 (1956).
149. Tabor, D., Winterton, R. H. S.: Proc. Roy. Soc. A **312**, 435 (1969).
150. Wittmann, F., Splittgerber, H., Ebert, K.: Z. Physik **245**, 354 (1971).
151. Hunklinger, S., Geisselmann, H., Arnold, W.: Rev. Sci. Instruments **45**, 584 (1972).
152. Israelachvili, J. N., Tabor, D.: Proc. Roy. Soc. (Lond.) A **331**, 19 (1972).
153. Bargeman, D.: J. Colloid Interface Sci. **40**, 344 (1972).
154. van Voorst Vader, F.: Proc. 6th International Congress on Surface Active Substances, Zürich, Sept. 1972.
155. Bailey, A. I., Daniels, H.: J. Phys. Chem. **77**, 501 (1973).

Interaction Constants

156. Davison, W. D.: J. Phys. B (Proc. Phys. Soc.) **1**, 597 (1968).
157. Gregory, J.: Adv. Colloid Interface Sci. **2**, 396 (1969).
158. Büttner, H., Gerlach, E.: Chem. Phys. Letters **5**, 91 (1970).
159. Langhoff, P. W.: Chem. Phys. Letters **12**, 217 (1971).
160. Langhoff, P. W.: Chem. Phys. Letters **12**, 223 (1971).
161. Pack, R. T.: Chem. Phys. Letters **13**, 205 (1973).
162. Gingell, D., Parsegian, V. A.: J. Theor. Biol. **36**, 41 (1972).
163. Krupp, H., Schnabel, W., Walter, G.: J. Colloid Interface Sci. **39**, 421 (1972).

Hydrogen, Inert Gases

164. Karplus, M., Kolker, H. J.: J. Chem. Phys. **41**, 3955 (1964).
165. Kolos, W., Wolniewicz: J. Chem. Phys. **43**, 2429 (1965).
166. Chan, Y. M., Dalgarno, A.: Mol. Phys. **9**, 349 (1965).
167. Kestner, N. R., Sinanoglu, O.: J. Chem. Phys. **45**, 194 (1966).
168. Kestner, N. R.: J. Chem. Phys. **45**, 208 (1966).
169. Getzin, P. M., Karplus, M.: J. Chem. Phys. **53**, 2100 (1970).
170. Fowler, R., Graben, H. W.: J. Chem. Phys. **56**, 1917 (1972).
171. Kolos, W., Les, A.: J. Quantum Chem. **6**, 1101 (1972).
172. Burke, P. G., Robb, W. D.: J. Phys. B: Atom. Molec. Phys. **5**, 44 (1972).
173. Robinson, G., March, N. H.: J. Phys. C: Solid State Phys. **5**, 2553 (1972).
174. Caligaris, R. E., Rodriguez, A. E.: J. Phys. B: Molec. Phys. **6**, L 24 (1973).
175. Björna, N.: J. Phys. B: Molec. Phys. **6**, 1412 (1973).
176. Sternheimer, R. M.: Phys. Rev. A **8**, 685 (1973).

Interaction Potentials, General

177. Hirschfelder, J. O.: J. Chem. Phys. **43**, 199 (1965).
178. Ohki, S., Fukuda, N.: J. Colloid Interface Sci. **27**, 208 (1968).
179. Kawatra, M. P., Lebedeff, S. A.: J. Chem. Phys. **51**, 4744 (1969).
180. Good, R. J., Hope, Ch. J.: J. Chem. Phys. **53**, 540 (1970).
181. Ninham, B. W., Parsegian, V. A.: Biophys. J. **10**, 646 (1970).
182. Richmond, P., Ninham, B. W.: J. Low Temp. Phys. **5**, 177 (1971).
183. Kihara, T., Yamazaki, K., Jhon, M. S., Khim, U. R.: Chem. Phys. Letters **9**, 62 (1971).
184. Arrighini, G. P., Biondi, F., Guidotti, C.: J. Chem. Phys. **55**, 4090 (1971).
185. Bargeman, D., van Voorst Vaader, F.: J. Electroanal. Chem. **37**, 45 (1972).
186. Allison, D. C. S., Burke, P. G., Robb, W. D.: J. Phys. B.: Atom. Molec. Phys. **5**, 1431 (1972).
187. Deb, B. M.: Rev. Mod. Phys. **45**, 22 (1973).
188. Robb, W. D.: J. Phys. B: Molec. Phys. **6**, 945 (1973).
189. Arrighini, G. P., Biondi, F., Guidotti, C.: Phys. Rev. A **8**, 577 (1973).
190. Gingell, D., Parsegian, V. A.: J. Colloid Interface Sc. **44**, 456 (1973).
191. Parsegian, V. A.: Ann. Rev. Biophys. Bioeng. **2**, 221 (1973).

Crystal Structure

192. Jansen, L.: Phys. Rev. **135**, A 1292 (1964).
193. Lombardi, E., Jansen, L.: Phys. Rev. **136**, A 1011 (1964).
194. Craig, D. P., Mason, R., Pauling, P., Santry, D. P.: Proc. Roy. Soc. A **286**, 98 (1965).
195. Jansen, L., Lombardi, E.: Disc. Faraday Soc. **40**, 78 (1965).
196. Pechhold, W.: Kolloid-Z. u. Z. Polymere **228**, 1 (1968).
197. Wobser, G., Blasenbrey, S.: Kolloid-Z. u. Z. Polymere **241**, 985 (1970).
198. Sangster, M. J. L.: J. Phys. Chem. Solids **34**, 355 (1973)

Prof. Dr. Dieter Langbein
Battelle-Institut e. V.
6000 Frankfurt am Main 90
Federal Republic of Germany

SPRINGER TRACTS
IN MODERN PHYSICS

Ergebnisse der exakten Naturwissenschaften

Atomic and Molecular Physics

Dettmann, K.: High Energy Treatment of Atomic Collisions (Vol. 58)

Donner, W., Süßmann, G.: Paramagnetische Felder am Kernort (Vol. 37)

Langbein, D.: Theory of Van der Waals Attraction (Vol. 72)

Racah, G.: Group Theory and Spectroscopy (Vol. 37)

Seiwert, R.: Unelastische Stöße zwischen angeregten und unangeregten Atomen (Vol. 47)

Zu Putlitz, G.: Determination of Nuclear Moments with Optical Double Resonance (Vol. 37)

Elementary Particle Physics

Current Algebra

Furlan, G., Paver, N., Verzegnassi, C.: Low Energy Theorems and Photo- and Electroproduction Near Threshold by Current Algebra (Vol. 62)

Gatto, R.: Cabibbo Angle and $SU_2 \times SU_2$ Breaking (Vol. 53)

Genz, H.: Local Properties of σ-Terms: A Review (Vol. 61)

Kleinert, H.: Baryon Current Solving SU (3) Charge-Current Algebra (Vol. 49)

Leutwyler, H.: Current Algebra and Lightlike Charges (Vol. 50)

Mendes, R. V., Ne'eman, Y.: Representations of the Local Current Algebra. A Constructional Approach (Vol. 60)

Müller, V. F.: Introduction to the Lagrangian Method (Vol. 50)

Pietschmann, H.: Introduction to the Method of Current Algebra (Vol. 50)

Pilkuhn, H.: Coupling Constants from PCAC (Vol. 55)

Pilkuhn, H.: S-Matrix Formulation of Current Algebra (Vol. 50)

Renner, B.: Current Algebra and Weak Interactions (Vol. 52)

Renner, B.: On the Problem of the Sigma Terms in Meson-Baryon Scattering. Comments on Recent Literature (Vol. 61)

Soloviev, L. D.: Symmetries and Current Algebras for Electromagnetic Interactions (Vol. 46)

Stech, B.: Nonleptonic Decays and Mass Differences of Hadrons (Vol. 50)

Stichel, P.: Current Algebra in the Framework of General Quantum Field Theory (Vol. 50)

Stichel, P.: Current Algebra and Renormalizable Field Theories (Vol. 50)

Stichel, P.: Introduction to Current Algebra (Vol. 50)

Verzegnassi, C.: Low Energy Photo and Electroproduction, Multipole Analysis by Current Algebra Commutators (Vol. 59)

Weinstein, M.: Chiral Symmetry. An Approach to the Study of the Strong Interactions (Vol. 60)

Electromagnetic Interactions

Deep Inelastic Lepton Scattering

Drees, J.: Deep Inelastic Electron-Nucleon Scattering (Vol. 60)

Landshoff, P. V.: Duality in Deep Inelastic Electroproduction (Vol. 62)

Llewellyn Smith, C. H.: Parton Models of Inelastic Lepton Scattering (Vol. 62)

Rittenberg, V.: Scaling in Deep Inelastic Scattering with Fixed Final States (Vol. 62)

Rubinstein, H. R.: Duality for Real and Virtual Photons (Vol. 62)

Rühl, W.: Application of Harmonic Analysis to Inelastic Electron-Proton Scattering (Vol. 57)

Experimental Techniques

Panofsky, W. K. H.: Experimental Techniques (Vol. 39)